SYSTEMS ANALYSIS
Definition, Process, and Design

SYSTEMS ANALYSIS
Definition, Process, and Design

SECOND EDITION

Philip C. Semprevivo
Deloitte Haskins & Sells

SRA®

SCIENCE RESEARCH ASSOCIATES, INC.
Chicago, Henley-on-Thames, Sydney, Toronto
A Subsidiary of IBM

Acquisition Editor	Terry Baransy
Project Editor	Ann Wood
Designer	Carol Harris
Compositor	Printed Page Graphics
Cover Photographer	Tom Tracy

Acknowledgments

The following photographs were reproduced with permission. Chapter opening, pages 31, 195, 305, 334–35, 345, 391, courtesy International Business Machines; pages 128, 154, courtesy TRW; page 260, courtesy of Kodak; page 335, courtesy of the Burroughs Corporation.

Library of Congress Cataloging in Publication Data

Semprevivo, Philip C.
  Systems analysis.

  Includes index.
  1. System analysis.   I. Title.
T57.6.S44 1982   003   82-3230
ISBN 0-574-21355-4   AACR2

10 9 8 7 6 5

To Peggy

Contents

Preface   xi

Introduction to Systems   2

What Is a System?   2
What Is Systems Analysis?   8

CHAPTER 1   Contemporary Systems Analysis   19

The Systems Development Life Cycle—An Overview   19
Implementing the Systems Development Life Cycle   32

CHAPTER 2   Effective Communications in Systems Analysis   49

An Approach to Effective Communications   51
Using the Interview   58
Interview Types and Techniques   61
Effective Group Communication   66
Effective Presentations   69
Improving Your Communication Skills   72

CHAPTER 3   Tools of the Analyst   80

Structured, Top-Down Approaches   80
The HIPO Technique   82
Data Flow Diagrams   88
Nassi-Shneiderman Charts   90
Standard Flowcharts   92
Decision Tables   98
The Questionnaire   103

CHAPTER 4   Problem Definition and Classification   118

Gaining Organizational Commitment   120
What Kind of Problem?   125
How Complex a Problem?   134

CHAPTER 5   Data Collection and Analysis   145

   Levels of Information   146
   Applying a Structured, Top-Down Approach   150
   What to Look For on the Job   153
   Recognizing Apparent Problems   156
   Methods for Analyzing Problems   161
   The Process of Documentation   169

CHAPTER 6   Systems Planning—Alternatives, Feasibility,   183
          and Proposal

   User and Management Involvement   184
   Planning Alternatives   185
   Other Design Considerations   199
   Systems Feasibility   200
   Selection of a Systems Plan   204
   The Systems Proposal   206

CHAPTER 7   System Cost Determination   213

   Systems Costs and Systems Benefits   214
   Cost Categories   217
   Comparative Cost Analysis   221
   Data Processing Costs   224
   The Data Processing Cost Center Concept   230

CHAPTER 8   Systems Design—A Structured Approach   238

   Structured, Top-Down Design   239
   Logical Design Requirements   246
   Data Administration and Data Dictionaries   261
   Auditable Systems   264
   Forms Requirements   268
   Forms Design   272
   CRT Screen Design   274
   Program Specifications   279
   Development Completion Schedule   281
   Structured Walkthroughs   281

CHAPTER 9   File and Data Base Organization and Design   291

   Management Information Systems (MIS)   292
   General Design Considerations   297
   File Organization   303
   Data Base Design   313

CHAPTER 10   Computer System Evaluation and Selection   327

The Need for Formal Evaluations   328
The Formal Evaluation and Selection Procedure   329
Factors to Be Evaluated   334
Price/Performance Factor   336
Systems Capability Factor   340
Systems Support Factor   343
Making the Selection   346
Minicomputers and Microcomputers   349

CHAPTER 11   Project Management and Control   357

The Development of Standards   362
Before Applying Standards   365
Project Control   368
Gantt Charts   372
The Critical Path Method   374
Program Evaluation and Review Technique (PERT)   382

CHAPTER 12   Systems Conversion and Implementation   389

Planning Considerations   389
Selecting a Conversion Method   396
Systems Follow-up   402
Quality Assurance of New Systems   403

**Appendix A   Training Program in Effective Communications
for Systems Analysts**   417

Appendix B   Data Processing Equipment Costs   420

Appendix C   Computer Evaluation and Selection   432

Index   439

Preface

The first edition of *Systems Analysis* introduced a broadened concept of the discipline—one located at the crossroads of technology, management science, and human relations. It also depicted a radically different role for the analyst: that is, rather than the traditional "all-knowing expert" the analyst was perceived as a catalytic change agent. Analyzing problems, exploring alternatives, confronting major policy issues and helping people to better understand technology were viewed as major aspects of the analyst's overall duties and responsibilities. Since writing the first edition, the importance of this broadened perspective has been reinforced by faculty, students, and practitioners. Because of this we have chosen to maintain, and in several instances, expand upon this approach.

Based on our recent survey of faculty who teach courses in systems analysis we have sought to maintain or enhance several components of the text. We continue to present educational objectives at the beginning of each chapter. End-of-chapter word lists and problems have been expanded. Because the course is increasingly taught to older students with prior work experiences, the work problems have been reoriented toward business—as opposed to educational situations. The end-of-chapter problems have been further enhanced by the introduction of a continuing minicase in which the student assumes the role of a systems analyst with Context International, Ltd.

Additional teaching aids include an expanded instructor's guide and a new student workbook. The workbook uses a case study approach with numerous practical work problems, questions, and student answer sheets. The workbook is particularly recommended for evening or continuing education students whose access to faculty for additional help may be limited.

Chapter 1 (as in the first edition) presents an overview of the systems development life cycle. However, it has been expanded in those areas dealing with organization, including a new section on the team model and its application to information systems development. Chapter 2 on effective communications has seen major enhancement. In addition to its coverage of interviewing and one-to-one interpersonal communications, major new sections on effective group communications and effective presentations have been developed.

Chapter 3 has likewise undergone major revision. A number of structured techniques have been described and illustrated including Hierarchy plus Input, Process and Output (HIPO) Data Flow diagrams, and Nassi-Shneiderman charts. Thus this chapter plays an important supportive role for subsequent chapters that deal with structured approaches to systems analysis and design.

Chapter 4, "Problem Definition and Classification," retains the well-received discussions of classical systems problems but now includes practical discussion on how to gain improved organizational commitment to the systems project.

Chapter 5, "Data Collection and Analysis," has been revised to be consistent with top-down, structured methodology. Although there are numerous writings on structured program design, this chapter is an early attempt to illustrate practically how to apply structured methodology during the early stages of analysis.

Chapter 6, "Systems Planning," has an expanded discussion of user and management participation during planning and of several technical alternatives, such as distributed data processing. Chapter 7, "System Cost Determination," continues to provide discussion on the broader range of systems costs and benefits of interest to organizations.

Chapter 8, "Systems Design," like Chapters 3 and 5, has undergone major revision to reflect structured, top-down approaches. Illustrative examples of structured charts and diagrams assist the student in understanding the application of structured methodology. The chapter has also been revised to reflect the growing prevalence of online transactional systems. CRT screen design and the structure and use of data dictionaries are representative of the extensions to be found in this chapter.

With the growing application of data base technology it was decided to expand Chapter 9 to include some discussion on each of the three major data base models: hierarchical, network, and relational. Like the retained discussions on file structures, these topics are presented at an appropriate introductory level.

Chapters 10, "Computer System Evaluation and Selection," and 11, "Project Management and Control," have modest revisions. These include a look at the evaluation and selection of minicomputers and microcomputers, as well as traditional mainframes. Also, a section on Program Evaluation and Review Techniques (PERT) was included in Chapter 11 to augment the discussions on Critical Path Method (CPM) and Gantt Charts at the request of a number of faculty who use the book.

Chapter 12, "Systems Conversion and Implementation," continues to stress the importance of planning for an effective systems implementation. The chapter has been further enhanced with a number of increasingly important operational considerations such as test system migration, security in an online environment, and data base backup and recovery methodology.

A great many people aided in the formation of this book. The following individuals provided useful comments and suggestions in the early stages of the development of the manuscript:

David R. Adams, Northern Kentucky University
Donald A. Davidson, Laguardia Community College
William K. Desilet, Duluth Area Vocational Technical Institute
Clinton P. Fuelling, Ball State University
Seth A. Hock, Columbus Technical Institute
James Hoey, Potomac State College
Paul W. Ross, Millersville State College
Alfred C. St. Onge, Springfield Technical Community College
Bruce M. Saulnier, Quinnipiac College
Jasper J. Sawatzky, Humboldt State University

Special thanks are due to Marilyn Bohl of IBM, William R. Cornette of Southwest Missouri State University, Alan L. Eliason of Eliason and Associates, and Carolyn M. Johnson of the MITRE Corporation, all of whom read the manuscript and helped immeasurably in perfecting the accuracy, clarity, and pedagogy of the book.

The second edition of Systems Analysis aims to maintain and enhance those aspects of the first edition that worked well for both student and instructor. This edition also introduces significant new materials based upon recent trends and developments. We have been fortunate in the past to receive excellent responses to our solicitations of faculty for feedback. We have also received, on occasion, unsolicited and valuable commentary. We hope that you will continue to assist us by responding to our requests for feedback and by directly presenting your thoughts and concerns to us.

Philip C. Semprevivo
New York, N.Y.

Introduction to Systems

CHAPTER OBJECTIVES

When you finish this introduction you will be able to—
- *Identify and define the characteristics of a system.*
- *Define systems analysis.*
- *Define the role of the systems analyst.*
- *Discuss the importance of understanding the structure of an organization in systems analysis, using examples of methods which may be used in gaining this understanding.*

The term *systems analysis* is one of the most frequently used buzz terms of modern business. Consequently, it is not surprising that there are many conceptions and misconceptions of exactly what systems analysis is.

The student who wants to work in the field as a professional systems analyst must meet at least two major requirements. First, he or she must gain a clear understanding of what systems analysis is, along with the duties and responsibilities of the systems analyst. Second, the student must learn to apply a variety of systems tools and techniques at appropriate points throughout the systems process.

WHAT IS A SYSTEM?

The word *system,* as it is used in systems analysis, does not necessarily mean either a computer or a group of computer programs. Although computing equipment and computer programs frequently play an important role in a system, they are not the only concerns of the systems analyst. The concept of a system is much more than this. That is, a system can be defined as a series of interrelated elements that perform some activity, function, or operation.

Let us take a moment and consider some of the different kinds of systems. Biological systems, of which the human body is an important example, are frequently discussed in high school and college biology classes. Likewise, we hear references to educational systems, social systems, business systems, and many others. Why can we define each of these very different examples as a system?

There are a number of common elements that can be found in each of the above examples:

1. Interaction with the environment
2. Purpose
3. Self-regulation
4. Self-correction

All of the systems mentioned thus far interact in some way with the world around them, which is referred to as their environment. It is this characteristic which results in their being called "open" systems. What this means, quite simply, is that the system will receive inputs from the environment and that it will produce outputs for the environment. It is through its inputs and outputs that a system is interactive with its environment.

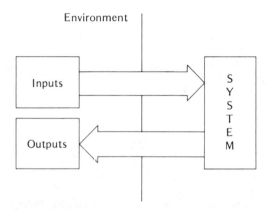

The second major characteristic of a system is that it has a purpose, goal, or objective. For example, a business system may have as its purpose a corporate goal, such as making a profit. The purpose of the human system is life.

A third major characteristic of a system is that it tends to maintain itself in a steady state. In this sense, systems are said to be self-regulating. Self-regulation is accomplished through the dynamic interaction of system component parts or subsystems; hence, self-regulation is internal. The manner in which the human body tends to maintain itself alive through the internal interactions of its component parts is a good example of a self-regulating system.

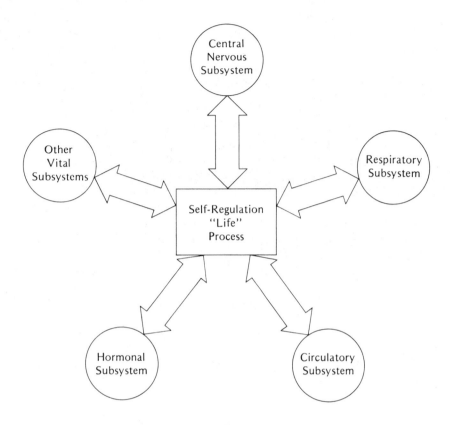

A fourth characteristic of systems is that they are self-correcting or self-adjusting. Many times, interaction with the environment results in conditions which upset the normal self-regulatory operation of a system. For example, interaction between the human body and the environment can lead to symptoms of the common cold. Under these conditions, if normal self-regulation is insufficient, the system must be capable of adjusting itself to the new conditions. In our example, the body must be capable of generating antibodies to fight off the cold. This self-adjusting capability of a system is partly explained through the concept of feedback.

At the time that a system creates an output, it samples that output and information about the effects of that output. This information is referred to as "feedback" since it is fed back through the system as inputs. The value of feedback is that it informs the system of required corrective action.

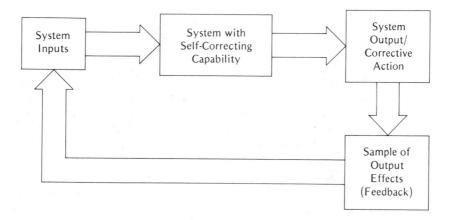

The self-regulating and self-corrective characteristics of systems are particularly important because they point out that systematic behavior is loop-like, not a straight-line progression of activities. As we analyze existing business systems and attempt to design new ones, we must remember these essential systems characteristics. Frequently, it is the lack of self-regulating and self-correcting mechanisms in the design of a system which leads to system failure later on.

Levels of Systems

A system is frequently made up of many smaller systems. It may also be a part of a larger system. Thus it is possible to talk about different levels of systems. Using our earlier example of the human body as a system, we find a number of smaller systems within it, such as the circulatory, respiratory, and hormonal systems. We call these *subsystems* within the human system. It is also possible to view the human body as a subsystem within a larger system called society. A subsystem is simply a system within a larger system.

When we begin to discuss the analysis of business organizations as a system, we will find that there are levels of business systems also. The particular level of system to be studied is a major factor in determining the complexity of the systems problem.

Business Systems

The focus of this book will be on one group of systems called business systems. A business system is defined as a series of interrelated elements that perform some business activity, business function, or business operation. The term *business,* like *systems,* should be viewed in the broadest sense and not restricted to private business corporations.

Business systems are found in many private and public agencies, including corporations, government agencies, hospitals, colleges, trade unions, and many others. While the specific problems, goals, and purposes of these institutions may vary, there are many similarities in the kinds of activities, functions, and operations that they perform and in the way they organize themselves to perform these jobs.

Let us take the payroll function as an example of a business system that can be found in any of the above mentioned business institutions. (See figure I-1.) We can identify this function as a business system since it does consist of a series of interrelated elements and because it is characterized by:

1. *Interaction with the environment*   There are inputs, such as time sheets indicating employees who worked, for how long, on what jobs, and at what pay rate. There are also outputs, such as paychecks, financial summary reports, and labor distribution reports.

2. *Purpose*   The system is designed to insure that employees receive the compensation which is due them according to the employee-employer contract or agreement.

3. *Self-regulation*   There are internal checks and balances which enable the system to function smoothly in a timely fashion so that payroll deadlines are met. For example, bottlenecks may be avoided by shifting payroll staff assignments according to the size of workload experienced at various points in the process.

4. *Self-correction*   The system is designed to check for unusual conditions and possible error. For example, there may be a system check for possible overpayment and a procedure for resolving such "problem" cases.

The payroll system may be viewed as a subsystem within the larger financial operation of a business. Also, it may consist of a number of smaller business activities or subsystems such as those involved in the preparation of payroll data for processing and making payroll adjustments and corrections.

It is worth mentioning at this point that, although there are similarities between different organizations in the kinds of systems

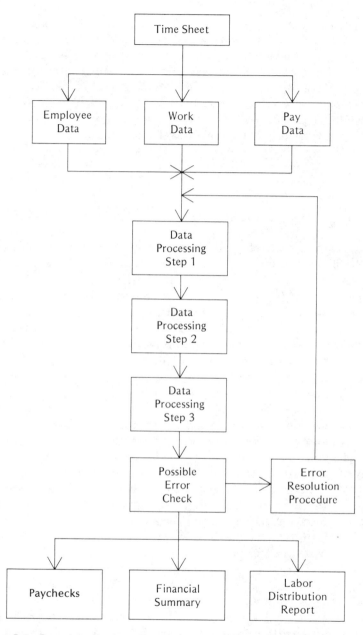

Figure I-1 Overview of a payroll system

they use, there are also elements unique to a given organization. These unique elements evolve from human and organizational preferences about the way work should be done. Thus, although two organizations may desire the same end product, such as a paycheck, one organization may believe in using time clocks, and the other may prefer an exception reporting system (that is, reporting only the time taken off from work and the reason for doing so). Furthermore, as a system's complexity increases, one can expect a corresponding increase in unique elements because of greater interaction among different groups of people in getting a job done. Thus, given two companies, we would expect to find greater differences between their total financial and accounting management-information systems than we would find if we simply looked at their payroll systems in isolation.

WHAT IS SYSTEMS ANALYSIS?

In our complex modern society we are not usually aware of the interrelationships that bind us together and seldom stop to think about them in the course of a typical day. For example, when we go out for dinner or to buy a new car, we remain relatively unaware of the way in which our actions affect the functioning of the economy. The fact that we create a demand for new automobiles or the need for additional jobs through our purchasing habits is rarely the focus of our attention or the reason why we act.

In a similar sense, the average businessman or woman usually does not base decisions upon a thorough understanding of the network of interrelationships and interdependencies which exist within the business organization. After all, it is difficult enough to deal with marketing problems without also becoming expert in production planning, accounts receivable, order processing, and the many other related activities which characterize the complexity of modern business. And yet, without some very real understanding of these interactions it is difficult to assess what the impact of decisions will be upon the total organization.

Systems analysis is the process of studying the network of interactions within an organization and assisting in the development of new and improved methods for performing necessary work. It is being increasingly applied to a wide variety of corporate, educational, and government institutions.

The systems analyst is a trained professional and therefore can assist an organization in planning for change. As a rule the analyst knows advanced problem-solving techniques, including the use of computers; but generally, the analyst's greatest contribution is to promote better understanding of problems and of the alternative solutions for dealing with them.

The Systems Perspective

The systems analyst views social, political, and economic institutions, such as business corporations, government agencies, colleges, and universities, as functional entities. The analyst assumes that these institutions exist to:

1. Perform some specific (and hopefully important) "jobs"
2. Perform these jobs in a way which can be understood through analysis
3. Perform these jobs for definite purposes (for example, to make a profit or to serve the public)

Furthermore, the analyst assumes that to operate effectively, an institution will organize its people and assign them specific tasks, resources, and responsibilities for ensuring that the important jobs are done and the institutional purposes fulfilled.

It is because organizations consist of people that they are dynamic and changing, not static or stable. They are characterized by a high degree of human interaction both within the organization and in their relationships to external groups, such as consumers or stockholders. Because they are so people-oriented, these dynamic organizations can experience many problems in trying to get a job done. Equally important, they offer opportunities for improvement in getting the job done through effective systems analysis.

The Role of the Systems Analyst

In many ways, systems analysis is a profession whose time has come. In the early days of data processing, when the focus was on automating small, self-contained business activities, there was only a limited need for extensive analysis. After all, only a few knowledgeable people were involved, and automation's overall impact on the total business was nearly transparent. At that time, the data processing personnel were primarily concerned with keeping abreast of the rapidly changing technology. They were busily engaged in writing (or learning to write) computer programs.

Today, all that has changed. With the growth of technology and the advent of large-scale, computer-based information systems, our thoughts about applying that technology have evolved from simple systems that a single individual could develop to large-scale systems that require extensive resources and many people to develop.[1] This transition has not been smooth, and it has shifted our attention from problems exclusively associated with computing hardware and software to a perspective concerned with people and organizational issues as well. In this broadened environment, the systems analyst, and not the computer programmer, is in the limelight.

To be effective, the systems analyst must achieve two primary objectives. First, it is necessary to observe, understand, and evaluate the interactions which routinely occur as a part of the job under investigation. This can be called the *assessment objective*. Normally, as a part of the assessment, it is important to answer the following questions:

1. What is being done?
2. Why is it being done?
3. Who is doing it?
4. How is it being done?
5. What are the major problems involved in doing it?

Secondly, the systems analyst offers specific suggestions for improving effectiveness in getting a job done. A primary goal of the analyst is to provide assistance by suggesting alternative solutions. Many times these suggestions involve modern technology such as computers and other advanced equipment. At such times the analyst may be viewed as a communications-link between technology and the user. To meet this *assistance objective* the analyst seeks answers to the following questions:

- What other ways exist for dealing with the problem?
- What kinds of benefits and liabilities are associated with these alternative approaches?
- What steps must be taken to develop and implement the chosen alternative?

The processes of assessment and assistance are not separate and distinct in their application. Rather, they are on-going and inter-related activities. As the analyst gains a better assessment of the problem, better assistance can be provided. On the other hand, knowing that certain kinds of assistance are available gives the analyst a broader perspective for true assessment.

To be effective in meeting both the assessment and assistance objectives, the systems analyst must establish good communications with all of the people affected. Because the analyst largely depends upon others for both information and cooperation, communications skills—not salesmanship and good ideas—are the essential ingredients in performing effective systems analysis. Chapter 2 of this text is dedicated solely to communications. For the present, we will say that communication occurs when there is a mutual understanding of what is said or written, both as stated and as implied. Communications skills are the tools and techniques which enhance the possibility for communication.

Understanding Organizational Structure

In attempting to get its work done, an institution organizes or reorganizes its people in various ways. When this process results in an "official" organizational structure, such as a department, bureau, or division, we consider this the "formal" organization. At other times, people are organized or organize themselves in "unofficial" ways to get a job done. This process creates what is called the "informal" organization. Not everyone in the institution is aware that the informal organization exists, because it is achieved through unwritten agreements and is not reflected in the institution's published organization chart. All institutions have both formal and informal organizations, and both types must be analyzed by the systems analyst.

The most prevalent type of formal organization is called a functional organization. That is, people are organized into units around the most important jobs or functions to be performed. For example, if an institution needed to perform personnel, sales, and production functions, it would—using a functional approach—organize its people around those functions by creating the necessary departments, bureaus, and divisions.

As a student you may have seen an organization chart for your college. In any event you are probably somewhat aware of the bureaucratic hierarchy of the college administration. Figure I-2 depicts the usual method of graphically representing such an organization. It is called a "military-style" organization chart since its primary concern is the formal patterns of rank, authority, and function.

From a systems point of view, the military chart has limited value since it fails to explain how the organization actually gets the job done. Therefore the systems analyst considers other ways of depicting the organization in order to obtain a complete picture.

Important organizational questions for which the analyst must find answers include:

1. What are the informal as well as the formal communication lines within the institution?

2. Are there key individuals/offices which should be focused upon in greater detail?

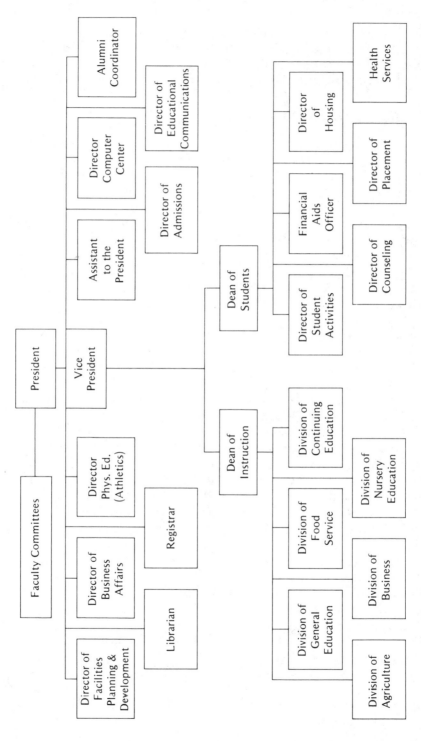

Figure I-2 Military-style organization chart

3. What interdepartmental and/or interagency cooperation and communication is essential to getting the job done?

4. What external people, groups, and institutions not covered by a diagram of internal staffing must be considered?

5. What changing patterns of authority and responsibility exist within the organization?

6. What political alignments exist and affect work procedures?

7. What rules (written and unwritten) affect the manner in which the organization functions?

Note that the military chart in figure I-2 does not include many important groups of people, such as students, faculty, parents, teacher unions, and taxpayers. Also, it makes no provision for the possibility that the Computer Center Director, the Registrar, and the Director of Business Affairs might have to routinely communicate with one another to get a job done (for example, registering students for a semester of college). Furthermore, the chart does not indicate that there will be times when personnel will deal directly with someone other than their immediate supervisor or subordinates. For example, there may be times when it is necessary and desirable for the Director of Computer Services to communicate directly with the chief administrative officer of the college. Finally, the chart fails to recognize the crucial importance of informal organization which exists and which is frequently guided by the political power and prestige of key individuals and the unwritten rules by which subordinates are expected to operate.

There are a number of methods for analyzing and representing an organization that can be used to fill the void between the military-style organization chart and reality. Comprehensive coverage of organizational analysis is generally treated as an advanced study topic in systems analysis. While the following comments on organizational analysis are brief, their brevity is more a function of the scope of this introductory text than of ultimate importance to systems analysis.

One important approach is to focus on the organization as it relates to specific functions. That is, what individuals and offices influence or guide the work? Using this approach the analyst determines whose opinion is sought when decisions must be made or problems resolved. By examining each major function, it is possible to identify key persons and offices who formally or informally affect what is being done and how it is done. These kinds of organizational relationships can be represented using simple flow charts (discussed in chapter 3) accompanied by functional descriptions.

Figure I-3 is a simplified example of how a flow chart and functional description can be used to illustrate organizational dynamics. In this example we see that the Vice President for Finance and his office are key elements and that their power and influence extend to other offices such as the computer center and the personnel office.

A second method of analyzing and representing organizations is conveniently referred to as *span of control*. Using this approach the analyst attempts to view each office or individual in terms of the number of subordinate people over which direct influence is exer-

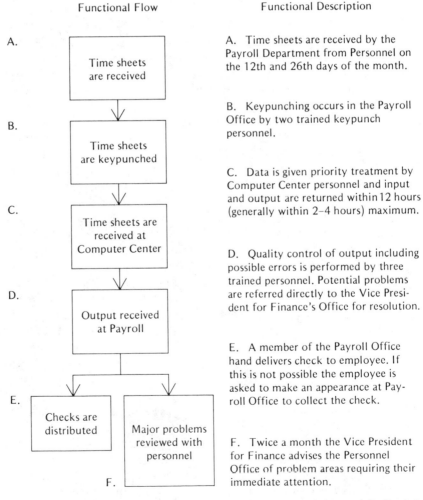

Functional Flow

A. Time sheets are received

B. Time sheets are keypunched

C. Time sheets are received at Computer Center

D. Output received at Payroll

E. Checks are distributed | Major problems reviewed with personnel

F.

Functional Description

A. Time sheets are received by the Payroll Department from Personnel on the 12th and 26th days of the month.

B. Keypunching occurs in the Payroll Office by two trained keypunch personnel.

C. Data is given priority treatment by Computer Center personnel and input and output are returned within 12 hours (generally within 2–4 hours) maximum.

D. Quality control of output including possible errors is performed by three trained personnel. Potential problems are referred directly to the Vice President for Finance's Office for resolution.

E. A member of the Payroll Office hand delivers check to employee. If this is not possible the employee is asked to make an appearance at Payroll Office to collect the check.

F. Twice a month the Vice President for Finance advises the Personnel Office of problem areas requiring their immediate attention.

Figure I-3 New systems implementation requires user training and feedback to work out problems

cised. Focusing on the number of people is critical to this approach since it views people as the major organizational resource; hence, a political and practical leverage for the administrative officer in

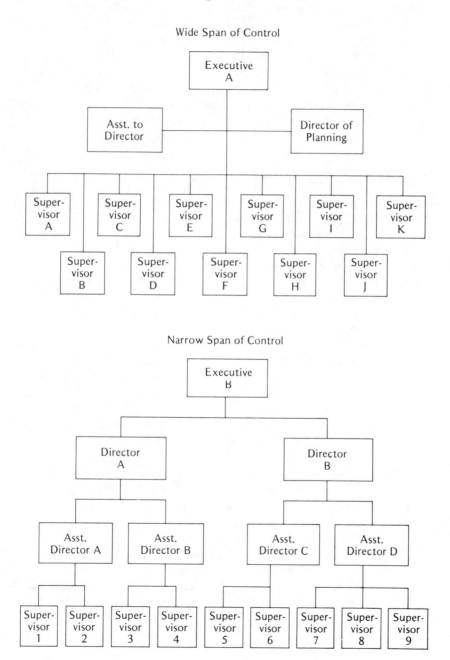

Figure I-4 Span of control

charge. When the span of control is wide, an individual will have many people upon which to draw for political and practical leverage. When the span of control is narrow, the organizational structure is characterized by more levels of hierarchy. Hence, direct control is exercised at each level over fewer people. (See figure I-4.)

Naturally there are other ways to view organizational units, including: (1) degree of control over change; (2) spheres of job influence beyond the boundaries of formal office responsibilities; and (3) personal factors such as friendship and other "connections." Each of these approaches can give the analyst valid insights into how an organization operates. Training and personal preference will largely determine which method(s) an analyst chooses.

Systems Technology and Organizations

As mentioned earlier, the most widely used form of organization is the functional organization. However, it is not the only alternative, nor is it always the most effective approach for dealing with complex systems that rely heavily on computing technology. The following brief example may help to illustrate this point.

Assume for the moment that we are managers in a medium-size manufacturing firm that is functionally organized. That is, there are a number of functionally based office organizations, such as personnel, payroll, accounting, and production planning. Let us also assume that our computer center people have proposed that we develop a new human resource system to provide more accurate and timely information about employees. The computer people argue that a single comprehensive system will provide numerous advantages to all offices.

For example, establishing a single system designed to serve the informational needs of all offices would be more efficient, because it would reduce the amount of redundant data stored on computer files. (If each office has its own system, each office must store and update much of the same information, such as employee name, address, and so on.) Also, there would be savings in the work load, because changes (such as address changes) by one office would correct that information for all offices simultaneously.

On the surface, these arguments appear compelling in support of developing a single comprehensive system. Yet, if we step back for a moment, we can clearly see a major incompatibility between the functional organization structure and the proposed cross-functional computer technology.[2] The offices simply were not organized to share information. Rather, they were established as quasi-independent functional entities.

This is not to say that comprehensive system concepts or functional organizations are good or bad per se. Rather, we want to point out that, in developing a new computerized system, it is important to view the system as something embracing both technology and the human organization to be served by that technology. Incompatibilities between the two can create difficulties for the systems analyst. Noting that these discrepancies exist, the analyst can begin to develop a systems strategy that permits these inherent difficulties to be overcome successfully.

NOTES

1. Kenneth T. Orr, *Structured Systems Development* (New York: Yourdon Press, 1977), p. 3.
2. Philip C. Semprevivo, *Teams: in Information Systems Development* (New York: Yourdon Press, 1980), pp. 8–11.

1

Contemporary Systems Analysis

CHAPTER OBJECTIVES

When you finish this chapter you will be able to—

- *Identify and describe the component parts of the systems development life cycle.*
- *List the major tools and techniques used in data collection and analysis.*
- *Describe the two points in the systems process where pilot or prototype systems are used.*
- *Discuss the program development responsibilities of the systems analyst.*
- *Name and discuss alternative methods for organizing the systems department within an institution.*
- *Discuss how a systems department ought to function to be an important asset to the people it services.*

THE SYSTEMS DEVELOPMENT LIFE CYCLE—AN OVERVIEW

Systems analysis in the broadest sense is a comprehensive process beginning with the definition of a problem and including the design and implementation of a new system which will correct the problem. This broad interpretation is assumed throughout this text.

The systems development life cycle contains the following activities:

1. Problem Definition
2. Data Collection and Analysis
3. Analysis of Systems Alternatives
4. Determination of Feasibility
5. Development of the Systems Proposal
6. Pilot or Prototype Systems Development
7. Systems Design
8. Program Development
9. Systems Implementation
10. Systems Review and Evaluation

A word of caution is given to the student at this time regarding use of the above list of systems activities. While it is true that we can, for learning purposes, isolate each of these activities, they are in real life highly interrelated. The systems analyst cannot expect that he or she will always encounter these activities in this order, or that once a task is finished that it will be finished for all time. To the contrary, you can expect a continual need to consider findings as tentative. Because they are tentative, these findings must be continuously tested and evaluated. Based upon the results, modifications can be made to the proposed solutions. Figure 1-1 illustrates to some extent how the systems development life cycle might be viewed interactively. Each of the important steps is explained briefly below.

Problem Definition

The systems analyst is seldom asked to find out if a problem exists. Rather, management determines that the problem exists and asks for help. The analyst will help management to assess the problem by determining its causes and assist in finding an effective solution.

Defining the exact nature of a problem is not an easy task and is frequently a time-consuming aspect of the total systems development life cycle. Management presentations of a problem are of little help from a systems point of view. For example, the management view of a problem might be, ". . . there is a high degree of customer complaint about the delivery of goods." From a systems point of view it is important to know if delivery service is negatively affected by the lack of available goods, slowness in processing customer orders, distribution procedures, or other causes. Many answers must be obtained—some from nonmanagement sources—before a clear systems definition of the problem can be determined. However, once the problem has been defined, the analyst will have not only a clear definition but a statement of objective for the remainder of the project (for example, "to reduce to 24 hours the time required to process a customer order to the point of notifying the warehouse of goods to be shipped.")

Data Collection and Data Analysis

The systems analyst is responsible for defining the problem clearly. In order to fulfill this responsibility it will be necessary to obtain the facts regarding the problem. This activity is generally called data collection and analysis. This does not mean that the analyst goes out and collects masses of information, brings it back to the office, and tries to sift through it for meaningful information. Rather, it means that the analyst will critically analyze data which is encountered to determine whether it is important to the task at hand.

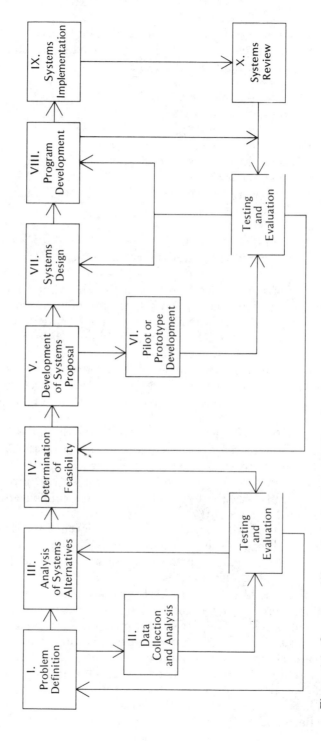

Figure 1-1 Overview of the systems development life cycle

There are a number of major tools and techniques for data collection and analysis. They will be discussed in chapters 2, 3, and 5 of this text and include:

1. Systems interviews
2. Standard flowcharts
3. Decision tables
4. Questionnaires
5. Flowcharts of the physical layout
6. Process flowcharts
7. Paperwork flowcharts
8. Written procedures
9. HIPO diagrams
10. Data flow diagrams
11. Nassi-Schneiderman charts

The primary point to remember at this time is that the systems analyst is dependent upon others for information, both about the problem and about the alternatives for solving it. The analyst must communicate effectively with individuals who are involved with the problem on a day-to-day basis.

Let us at this point attempt to illustrate problem definition and data collection as they might relate to a process with which you have had some experience, namely, admission to a college.

In our example let us say that the college administration at Golden University has become alarmed over the large number of complaints from prospective students and their parents regarding delays in the notification of acceptance. The problem at Golden University is accented by the fact that college enrollments have dropped in each of the last three years.

A systems analyst will begin by collecting and analyzing data to get a clear definition of the problem. The data search will focus on the college's responsiveness to prospective students' applications during the admissions process.

The initial thrust will be to investigate and fully document the manner in which the college currently processes an admissions application. The analyst will interview personnel in the admissions office and other offices involved in the admissions process. Figure 1-2 represents the formal organization of the Admissions Office at Golden U. There are four organizational units: recruitment, application receiving, application evaluation, and notification of students. Throughout the data collection, the analyst will document where and when in the process problems occur—particularly those

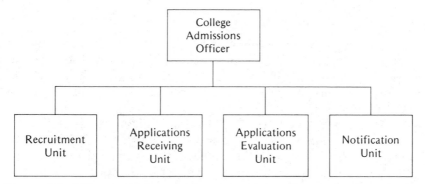

Figure 1-2 Military-style organization chart for the admissions office at Golden University

which cause delays. With fully documented observations, the analyst will be able to explain to others what the true systems problems are.

In the above example let us say that the following problem areas were found.

1. Delays in the evaluation of the applicant's prior academic record. These were caused by: (a) poor procedures for obtaining records from high schools and previous colleges attended; (b) poor procedures for collating prior academic records into the application folders; and (c) poor procedures for coordinating work flow between the Applications Evaluation Unit and the Applications Receiving Unit, which are located on different floors of the same building.

2. Delays in notifying a student of acceptance or rejection. These were caused in the Notification Unit after evaluation of the application had occurred by a cumbersome procedure which required that: (a) each student's folder be pulled and manually updated with the latest admissions decisions; (b) a data processing form be filled out after step "a" and sent to the college computer center; (c) official notification of processing by the computer center be received prior to any further action; and (d) an individualized form letter be prepared and hand signed by the admissions officer.

As we continue to outline the systems process in this chapter, we will frequently return to this College Admissions System illustration and elaborate upon it.

Analysis of Systems Alternatives

The complexity of modern business precludes the possibility of finding any one solution to a major problem which will be ideal in all respects. Generally it is a matter of weighing the relative advantages and disadvantages of several alternatives before deciding which system is "best."

To approach problem solving in a hard-sales fashion as if there were a single ideal system is both unwise and unprofessional. The analyst should not try to tell people what they must do, but outline what can be done, at what expense, and with what potential impact. The final decision as to which alternative is best should be made by management and by those who must live with the resulting system.

Acquiring skill and knowledge of systems analysis will not make you an all-knowing expert. The longer you work in the systems profession, the more you will come to understand the limits of your ability to solve, singlehandedly, other people's problems for them.

Recognizing the need for giving others a choice between several alternatives is an important ingredient in systems analysis. It is wise to begin early in building cases for several helpful alternative courses of action.

Returning to Golden University's College Admissions System, let us examine how the analyst might propose systems alternatives. One frequently used strategy is the "task force," or "team" approach. A representative group is established to review specific problems and develop new methods for dealing with them. In this instance the team might include the systems analyst, a student who has recently been admitted to the college, a representative from the admissions office, a faculty member who has recently served as an advisor for first-year students, and a representative from the college management.

For purposes of simplicity let us look at only the second of the two major problems previously noted and see what alternatives may exist. The cumbersome procedure for notifying a student of admission to the college might reasonably be changed in a number of ways:

Solution 1
- The Application Evaluation Unit will notify the computer center directly of its decision rather than delaying that process until the Notification Unit has had a chance to update the individual student folders.
- A less personalized letter which does not require the signature of the admissions officer will be used.

Solution 2
- Computer terminals will be used as follows:
    a. The Application Evaluation Unit will use these terminals to update computer files directly at the time that a decision is made.
    b. The Notification Unit will use these terminals to check on the status of applicants. The application folders will no longer be manually updated.
- The computer will be used to automatically generate a letter of acceptance or rejection based upon the decision it receives from the Application Evaluation Unit. The letter will be sent from the center to the mail room with a copy to the Notification Unit to be filed in the applicant's folder.

The above two alternatives illustrate two extremes for dealing with the problem. Solution 1 represents a plan for improvement upon what is currently being done with little technical change. Solution 2 on the other hand is a proposal for dramatic change and as such is much more of a new systems proposal. Actually the team would probably consider many "middle road" solutions incorporating some blend of ideas from both Solutions 1 and 2.

Determination of Feasibility

For each alternative that attempts to deal with the problem, it is necessary to conduct a series of tests which will establish clearly the benefits and the liabilities of the proposed plan. After subjecting each alternative to close analysis, one can base the ultimate decision upon the best information available.

Many people confuse making a systems proposal with conducting a feasibility study. Developing a systems proposal does not, as we shall see, automatically mean that alternative plans of action were considered or that the proposal was subjected to the rigors of feasibility testing.

Feasibility studies generally focus on three important questions:

1. What are the demonstrable needs and how does the proposed system meet them?
2. What capacity (short- and long-term commitment of resources) is required by the proposal and is available?
3. What is the anticipated impact of the proposed system upon the organization and what evidence is there of the organization's ability to react positively?

Satisfactory answers must be obtained in each of these areas if there is to be any assurance that the system will have a chance for

VISTA

success. The high risk of by-passing this important aspect of analysis has been demonstrated many times.

The work of the systems team at Golden University may well have included determination of feasibility for their two alternative solutions. In order to do this Solutions 1 and 2 should be tested as follows:

1. To what extent does the proposed solution speed up the notification process? For example, how long will it take to perform the notification process from receipt of the application?

2. Are the benefits (of quicker notification) sufficient? still too slow? faster than truly needed?

3. Are there other benefits associated with the solution not directly related to the problem of responsiveness but worthy of mention? For example, decreased admissions staff requirements.

4. What additional resources, such as equipment, additional personnel, and retraining of existing personnel, are required to implement the solution? For example, additional computer terminal costs.

5. Are the additional resources available? Do they cost too much? Are they easily financed?

6. Will the new system tend to have positive or negative effects on employee morale, productivity, and so on?

7. Is the current organization equipped to perform the work and make the necessary decisions required in the new system?

8. Is the current organization technically capable of designing and implementing the kind of system being proposed?

With the answers to these and other questions, the team at Golden University will be better able to select an alternative which will meet the university's needs and have a high degree of probable success. The results of feasibility determination by the team will then be documented and included in a formal systems proposal as indicated in the following section.

Development of the Systems Proposal

If the systems analyst is to gain the full support of everyone affected by the proposed system, then all parties must understand what is being proposed as a problem solution. To communicate both the nature and the scope of the proposed solution, a formal document (systems proposal) is prepared. This proposal is reviewed by each party, modified as required, and formally agreed to. In effect, it is a formal contract between parties and serves as a guideline for the duration of the systems project.

Any modification to the proposal must be made in writing and agreed to by all parties affected. While this much formality may seem strange, consider the size, complexity, and cost which may be associated with a major systems project. Certainly good planning and mutual understanding and agreement is essential from the beginning. It is simply good common sense to clarify, in writing, each of these systems requirements before committing to do a job.

Pilot or Prototype Systems Development

Pilot or prototype systems are commonly used at two different points in the systems process. First, there are times when a proposed solution is so novel that people are not sure that it will, in fact, work. It may be a solution that has never been tried before. Or it may involve methods and procedures that are radically different from those currently used within the organization. In such cases it is wise to develop the proposed solution on a limited scale before committing to a complete systems design.

A second use for pilot systems is during the implementation of a new system. In order to minimize the impact of systems implementation, a pilot or limited version of the complete system can be used. The pilot approach lets the organization test the new system and make required modifications without completely disrupting the day-to-day business process.

Imagine for a moment that the final admissions system proposal at Golden U. included the recommendation to use computer terminals. Let us also say that the college computer personnel had no prior experience in this kind of application.

Given the above circumstances, it would be wise for the team to recommend the development of a pilot system that would demonstrate the feasibility of the proposal. Furthermore, the recommendation might explicitly limit the pilot system to a demonstration that using the college computer it is possible to update and retrieve data files through the use of computer terminals. In addition to proving feasibility, a pilot system of this kind would provide the opportunity for computer personnel to gain the knowledge and experience which they currently do not possess, but which they will need to design and implement the computerized aspects of the system smoothly and efficiently.

Systems Design

The term "systems design" describes both a final product and the process by which the product is produced. In the design process, the analyst takes the necessary steps to ensure that the system will function as a system. There are generally two phases to the design process, namely, the logical design phase and the physical design phase.

During the logical design phase, or, as it is sometimes called, the "macro" design phase, the analyst focuses on developing general design specifications for the system. Logical design specifications include the identification of all desired outputs, such as management reports as well as a description of desired systems functions, in terms that relate to the business environment. Thus, for example, the kinds of data to be stored are identified in the logical design, but the exact way that those data will be organized and stored in the computer are not specified.

At times, the physical design phase is also called the "technical" or "micro" design phase. Having previously identified the logical design requirements, during this phase the analyst describes in very specific terms the various systems components, including the required outputs, inputs, and processes. The characteristics of the information base are also specified, and the procedures for operating the system are clearly delineated. When the design calls for computing equipment, a scheme for using that equipment is developed, and computer programming requirements are identified in general terms. These computer programs are specified in greater detail during the next phase, namely, program development.

The fruits of careful prior systems analysis are realized during design. If the earlier stages of analysis have been professionally conducted, the analyst understands the informational requirements and has the organizational commitment necessary to succeed in the two design phases. Consequently, the resulting design will be a truly effective master plan.

At Golden U., the team approach was also used for systems design. The first team action was to develop a plan that detailed a project schedule for (a) completing the design, (b) stating the new system's objectives, (c) determining criteria for judging the system's success, and (d) assigning persons to work with the team on various design aspects. A commitment to this plan was obtained from top university management before beginning the actual design.

During the logical design phase, the team was careful to identify all the major systems requirements before starting on specifics. Thus, at the conclusion of this phase, the team had completed a structural overview of the new system. The team could then perform its next task—technical design—without having to debate whether or not the system would be expected to perform a particular function.

During the second design phase, the system's components were described in great detail. Flowcharts were prepared, outlining the new admissions process, data file and record layouts were developed, sample forms and reports were designed, general programming requirements were presented, and user and computer center instructions and procedures were constructed. The results of the design process were organized as a cohesive design document and shared broadly with management and user personnel who had not directly played a role in its development.

Because Golden U. had done a good job in documenting its existing system and because a careful study of feasibility was made, its systems design effort was in effect a continuation and updating of what had already been documented. Minimal time was lost because an answer was not readily available or because someone had to construct a form a second time.

Program Development

Ideally, the systems analyst has no computer programming duties. It is difficult enough to do a good job with the systems aspects of a problem without also trying to code, debug, and test computer programs. The primary responsibility of the systems analyst should be systems analysis. Program development responsibilities of the analyst generally include:

1. Providing the programmer(s) with detailed specifications which clearly outline systems requirements

2. Responding to requests of programmers for additional information or problem clarification

3. Acting as an interface between computer programming efforts and user concerns

4. Establishing a basis for testing the accuracy and reliability of new programs

5. Maintaining contact with programming management to ensure that programs are completed in order and on time

At Golden U. the systems analyst decided to approach the problem of interfacing with the computer programming staff in the following way. In an initial meeting with the manager of computer programming the analyst asked that they work together to develop a project completion schedule and a systematic procedure for reporting on the project status. Also, he gained the manager's approval to present a systems overview to the programmers who would be assigned to the project. While the programmers would each be working on one or two independent segments of the system, the analyst reasoned that it would be advantageous if the programmers understood how their work related to the total project. In that meeting he also encouraged each of them to contact him if and when problems or ambiguities arose.

Systems Implementation

A new system must be implemented or placed into use in a way that minimizes confusion and general disruption of the day-to-day process of business. Implementation of a new system is frequently complicated by the fact that an old system is already in use. At such times it is necessary to convert or change from the old system to the new system. It is also possible that the new system will require additional new equipment. Careful planning, effective training of personnel, and a systematic implementation procedure are three important methods for dealing with systems implementation and with the complications which sometimes result from systems conversion and the introduction of new equipment.

At Golden U. the team felt that the dramatic changes proposed in the new system required an implementation plan that would safeguard against new system problems or "bugs." Accordingly it was decided to implement the new system for the Spring semester (when new applicants would be few) and to do so utilizing a "parallel" systems approach.

The parallel systems approach meant that the college would operate the old and the new system side-by-side for one semester. This would give personnel one semester to become accustomed to the new procedures and to iron out problems.

While the cost of operating two systems would be higher, the college felt that the safety factor was well worth the added cost.

Systems Review and Evaluation

There is no foolproof guarantee that problems will not occur during the initial months of new system use. When problems do occur, user personnel may not understand how to cope with them. Therefore, it is important to plan for a continuing support service. However, the analyst should establish some cutoff point after which contin-

Systems implementation

uous system support cannot be assumed. Naturally, a cutoff point cannot be arbitrarily decided. Rather, some period of time wherein the system performs acceptably is generally agreed to beforehand by all parties. Unless there is some justifiable reason for continuing the full-time commitment of professional staff beyond that point, continuous system support is stopped.

At Golden U. the systems team was reassembled for a post-implementation review. Comparisons were made to determine the extent to which the new system adequately met the mandates of the initial systems proposal. Also the team reviewed specific problems which were presented to it by the users.

In several instances the team directed data processing personnel to make specific changes and corrections which it deemed to be both necessary and feasible. However, in two instances the team concluded that user requests exceeded the limits of the initial commitment and proposed them as agenda items for future systems consideration. This meant that the value of making these changes would have to be reviewed in the light of other important requests for new systems development.

One important decision of the post-implementation team was to establish a cutoff point for the receipt of requests to make changes to the system. At the conclusion of the review and evaluation, requests for changes were referred to management as "new" requests. The personnel who worked on the admissions system were thereby freed to turn their full attention to new problem areas.

IMPLEMENTING THE SYSTEMS DEVELOPMENT LIFE CYCLE

The fundamental question posed in this section is, "How does systems analysis fit into the organizational structure of an institution?" There are two aspects to this question:

1. What is the relationship of systems personnel to the management of an organization?
2. What is the relationship of systems personnel to other employees?

Here, we are not talking about how a particular analyst works with other people, such as clerks, supervisors, and managers, during a given systems project. At this point we are simply looking at systems personnel, in general, and how they are organized to meet the systems needs of a particular institution.

Centralization Versus Decentralization

Many institutions have some form of centralized department that includes systems analysis. Since the titles of these departments vary considerably, it is not uncommon to find names like Bureau of Management Improvement, Office of Computer Systems Development, Department of Management Information Systems, and many others, each of which refers to a centralized group of systems personnel.

It is also common to find systems analysis organized as one or more units within a data processing department. Because the development of new systems is usually paralleled by the development of specific computer programs, these organizational units are frequently referred to as the Systems and Programming Unit.

In some very large organizations, the systems function is decentralized. That is, systems analysts are permanently assigned to business areas such as production planning and manufacturing.

Because centralization of systems personnel is common practice, most of our examples will refer to that kind of organizational structure.

What Are the Standards for a Systems Department?

There is no single, generally accepted method for structuring the relationship of systems personnel to either management or the other departments within an organization. Instead, one can find a full range of authority and responsibility in systems departments in different companies and agencies. In some organizations, systems personnel have management-like authority, but in others they are the pawns of every department chief in the company.

From our point of view, it is important to understand how different systems departments actually function. But it is also important to consider how a systems department *ought* to function to be an important asset to the people it services. Let us consider the ideal case first.

The Systems Department ought to establish an effective relationship with both management and nonmanagement personnel. To facilitate this, we would expect the organizational arrangement to encourage the following standards within the Systems Department:

1. The proper ordering of priorities within the Systems Department—the analysts should spend most of their time and energy on the most crucial problems.

2. Maximum utilization of resources—people and equipment should do primarily those things which they are best able to do.

3. Proper planning within the Systems Department—the analysts should construct realistic short-term and long-term plans, which can then be implemented.

4. Creation of a working atmosphere which promotes timeliness in response to changing conditions—the Systems Department should be permitted the flexibility required to work in a field which is characterized by rapid change.

5. Development of good communications between systems personnel and other people within the organization.

If there were a known method of organizing systems personnel that could ensure these five standards, we would have to agree that it was, indeed, a very excellent method of organization. But living as we do in a world of realities, what we find in practice is less than ideal. Sometimes the shortcomings are in the method of organization itself; at other times they are in the way people carry out or fail to carry out their assigned responsibilities.

Because the relative placement of systems personnel within an organization can have a dramatic effect upon the overall ability of analysts to do their jobs, it is important to examine some of the real alternatives. Hopefully, upon examining a limited number of hypothetical examples (organization models), you will begin to appreciate the wide range of possibilities. An awareness of their potential benefits and difficulties should stimulate your thinking about how *you* might, under similar conditions, work to improve your effectiveness as a systems analyst. The organization models which will be discussed include:

- The Systems Initiation Model
- The Steering Committee Model
- The Systems Clearing House Model
- The Corporate Planning Model
- The Service Center Model
- The Team Model

The Systems Initiation Model

In this model, the Systems Department itself initiates the development of new systems based upon the exposure and experience of its members to what has been done or is being done elsewhere. A partial organization chart in figure 1-3 illustrates this relationship. Generally, the Systems Department can do this only if it has considerable sanction from top management, and hence, a well-established line of communication to that management. In this kind of organizational approach, systems managers spend most of their efforts developing new proposals and selling them to management.

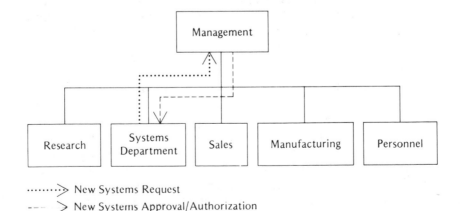

........> New Systems Request

--- > New Systems Approval/Authorization

Figure 1-3   Systems initiation model

There are some very real advantages to the Systems Department under this approach, beyond the rather apparent political niceties. For one thing, maximum use is generally made of existing personnel since projects are proposed for which the staff is well-trained and experienced. As a side benefit, this kind of organization contributes to the morale of systems employees, who feel secure and to some extent able to influence their destiny. Also, resource planning is improved since the Systems Department is free to set its own priorities and time schedules.

Notwithstanding the flattering position which the practice of this model yields to the Systems Department, many pitfalls can be associated with it. For example, although priorities are readily assigned, they are frequently assigned in a manner which is not beneficial to the true needs of the organization. Similarly, planning, although efficient, tends to be short-ranged or even near-sighted. The broader understanding of what should be done by a company, college, or government agency is not always apparent at the Systems Department level.

Perhaps most importantly, a strong systems initiative often causes problems between systems personnel and other members of an institution. The practice of this model has traditionally led to mistrust of the aims and motives of systems personnel. This lack of trust can break down the effectiveness with which otherwise capable systems professionals are able to do their jobs.

It is probably true that few organizations have consciously planned to employ this model. Rather, the practice has developed as the result of an aggressive and presumably capable individual or group within the Systems Department. It has probably continued because of specific projects which have "made a name" for the institution as progressive or even as a leader in its application and use of automation.

The Steering Committee Model

A steering committee usually consists of representatives from a number of different business areas. For example, in a private corporation a steering committee might consist of representatives from:

• Top management—for example, an executive vice president
• Middle management—such as representatives from marketing, manufacturing, production planning
• Technical managers—representatives from systems, data processing, research and development, and the like

It is the purpose of this committee to review and act upon all requests and proposals for systems analysis assistance. Requests are reviewed with the intent to:

1. Maximize the use of systems resources toward the accomplishment of overall organizational purposes and goals.

2. Establish priorities for systems study areas and the projects to be considered within these areas.

3. Provide maximum commitment of organizational support to projects which are undertaken.

4. Improve short-range and long-range plans for the use of systems resources.

A partial organization chart in figure 1-4 shows organizational relationships using the steering committee model.

Although this approach to the organization of systems analysis tries to bind together both technical and practical information, it does not guarantee success. Several pitfalls do exist. For example, the steering committee, like most committees, can be clumsy as a decision-making body. While committee members appear to agree inside the meeting room, they frequently lose identification with committee goals and purposes upon returning to their own offices. Furthermore, committee decisions are often affected by the vested interests of small groups or cliques within the committee.

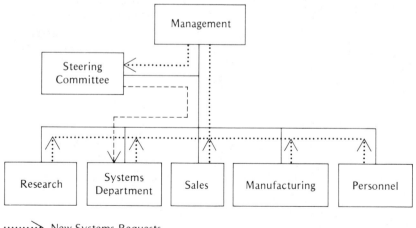

····> New Systems Requests
----> New Systems Approval/Authorization

Figure 1-4   Steering committee model

Still other pitfalls can be associated with a steering committee as with any high-level decision-making body. For example, unless care is taken to promote vital input from lower ranking personnel who work with existing problems on a day-to-day basis, the committee may tend to be somewhat unrealistic in terms of its expectations. Failure to recognize the operational facts of life can lead to decisions which attempt the impossible.

Last, there remains the distinct possibility that the steering committee may be simply a reactive body—not a leadership body, and that its major concern is to maintain the status quo. Certainly, this is a risk whenever responsibility is centralized within an organization. In these instances the steering committee may become a focal point for applying pressure and degenerate into a political tool rather than a business one.

The Systems Clearing House Model

This model attempts to place considerable, but not all, decision-making responsibility within the Systems Department. It assumes

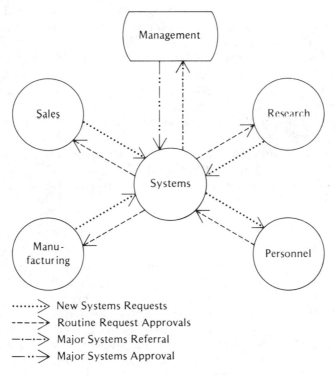

Figure 1-5 Systems clearing house model

that there is a certain level of request for routine systems services about which the Systems Department can make final decisions. On the other hand, the Systems Department will frequently refer new requests to higher management, as in the case of requests for major new developments. Figure 1-5 illustrates how the systems clearing house might operate.

This model may be viewed as a middle-of-the-road approach. In this sense it is a somewhat flexible approach with several distinct advantages. The Systems Department can react directly to the immediate routine needs of people in other departments without the usual red tape of gaining management approval. Also, the fact that the Systems Department contributes to, but does not actually make, the final decisions on important new systems requests is significant. In this way, final decisions will reflect the views and opinions of top management. This helps to eliminate many of the negative feelings about systems personnel which are common with the systems initiation model.

However, the Systems Clearing House Model is difficult to apply satisfactorily. Our statement that the Systems Department will make some decisions and top management will make others sounds many times more flexible than it is. Imagine for a moment that you are a systems analyst giving routine assistance to the Accounting Department. Imagine also that Accounting Department personnel feel your service is vital. Suddenly, management decides that the company will undertake development of a major new order-processing system. At this point, what happens to the commitment of the Systems Department to Accounting? In practice, the routine commitment is usually "suspended" for the duration of the new systems commitment.

This kind of on-again/off-again attention to people's immediate needs characterizes most attempts to use this model. As a result, users lose confidence in the Systems Department's reliability.

The Corporate Planning Model

Growth of the multiple-company concept in private industry and the increasing centralization of government have led to the practice of establishing systems groups at a fairly high level within some organizations. These systems groups serve as a primary tool of management in the evaluation of operational efficiency and as a task force in attacking specific problems diagnosed by management. In this manner, the Systems Department is effectively removed from the normal chain of command of the organization and employed as a "brain trust" of technology and procedure.

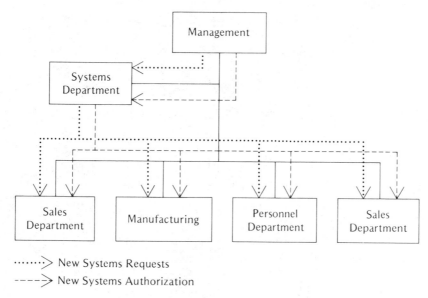

> ······> New Systems Requests
> ----> New Systems Authorization

Figure 1-6   Corporate planning model

The partial organization chart in figure 1-6 shows how, for new systems, there is a flowing down of requests and authorization from the top through the Systems Department quite outside of the mechanics of the normal organization.

This is a difficult concept for an analyst to assess objectively, primarily because it provides an interesting, responsible, and well-paying job for the individual analyst. However, because the systems group is removed from the mainstream of bureaucracy, friction can easily develop between systems personnel and other departments within the organization. Middle managers in particular often feel that their opinions are ignored; yet they are left with getting the job done. From their point of view, priorities are set by the analyst to please the whims of top management rather than to meet the requirements of everyday operations.

This kind of model is not widely employed, and the requirements for obtaining this kind of systems position are very high.

The Service Center Model

The Service Center Model is employed by large corporations that have decentralized their operations into cost/responsibility centers. Each of these centers has considerable autonomy but is evaluated by management in terms of its overall performance (that is, profitability).

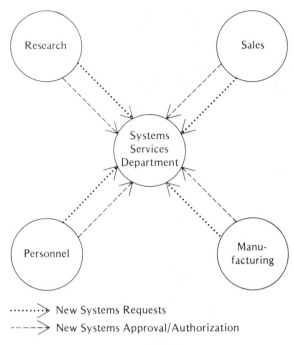

Figure 1-7   Service center model

A Data Processing or Systems Service Center is usually established by the company and may be contracted for use by any of the cost/responsibility centers. While the Systems Department does not always have to show a profit, it is usually expected to cover its operating cost.

This approach offers an interesting contrast to the steering committee concept. Rather than deal with a centralized group of user and management personnel, the Systems Department must now negotiate individually with each separate user for each proposed systems project.

Figure 1-7 illustrates how the Systems Department receives both requests and final authorization for individual system projects.

Many users who have worked with this kind of organizational arrangement are favorably impressed. In particular, they seem to appreciate having greater say about what gets done for them and when. A company using this approach often allows cost/responsibility centers to buy services outside of the company if the Systems Service Center cannot deliver the right product, at the right time, and for the right price.

The Team Model

The discussion of the Team Model has been saved for last, because its use does not preclude the use of any of the preceding models. In fact, it is very probable that another model, such as the Steering Committee Model, will be used to make major decisions about which systems to develop, and the Team Model will be used as the vehicle for implementing the systems development life cycle for any given project.

The systems development team, as shown in the partial organization chart in figure 1-8, consists of a representative group of people from top management, the Systems Department, and the major units being affected by the new system. The team is a work group responsible for developing the system or a particular phase of it, such as systems design. The team is usually composed of only five to seven full-time members, but it can be expanded with additional part-time members if people with special expertise are needed by the team at some point in its work. For example, if the team were engaged in technical data base design, it would want to add a part-time data base specialist for that aspect of the work.

Selecting team members is a particularly critical event, because the quality of team membership is directly related to the quality of

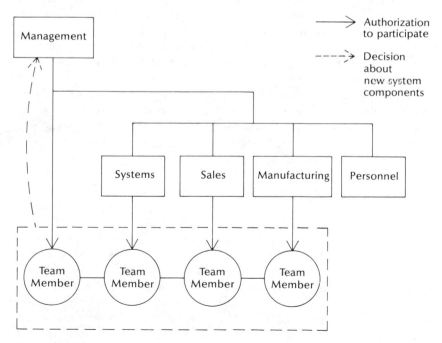

Figure 1-8 Team model

the final product. It is important to obtain the best possible people to work on the team. They should be bright, dependable, and hardworking, and they should possess the interpersonal skills necessary to work well with others. Also, they should possess the authority to make commitments and decisions on their own, and they should be generally well respected by the management and the user communities.

The Team Model has become a predominant approach to implementing new systems, particularly large-scale systems that cut across an organization's functional lines. This approach maximizes the opportunity for management and users to participate directly in systems development. Consequently, this approach provides an organizational bridge between the functional organization and crossfunctional computer technology.[1]

CHAPTER SUMMARY

In this chapter the student was provided with background information on the nature of systems and systems analysis. Using the unique systems view of an institution, the systems analyst is able to accurately assess institutional problems and performs an important service in assisting others to find effective problem solutions.

Systems analysis is given a broad interpretation, which includes the total life cycle of defining a problem, developing an effective solution for the problem, and implementing that problem solution as a new system.

The effectiveness with which a systems analyst is able to perform systems work is influenced by his or her skill as an analyst. Also, many other factors, such as management objectives and constraints and the manner in which the systems function is organized within an institution will have important consequences for the work of the analyst.

There are a wide variety of ways in which systems people are organized in different institutions. The effectiveness with which a systems analyst does a job can be greatly affected by the relationships to management and non-management personnel that are dictated by these formal organizations. In this text we want to stress that the user should be encouraged to take an active role throughout the systems process. Consequently, it follows that the systems analyst is responsible for finding ways which encourage user participation within the context of the existing formal organization.

Regardless of the particular style of organization, the role of management as expressed in organizational objectives and constraints will prove to be a major factor in what the analyst can and cannot do. For example, if a new and vigorous management is seeking as a goal the total reorganization of the institution, the ensuing systems analysis will be very comprehensive. In effect the analyst will have carte blanche to conduct a thorough and complete analysis. If, on the other hand, management is extremely satisfied with the fundamental business operations, systems analysis may take the form of suggestions for very specific operational improvement or even "window dressing."

Taken as a whole then, it will be some combination of management objectives and constraints coupled with existing relationships between the system department and management and non-management personnel which will influence the manner and effectiveness of the systems effort.

WORD LIST

business system—a series of interrelated elements that perform some business activity, business function, or business operation

communication—a process that occurs when there is a mutual understanding between parties of what is said or written both as stated and as implied

cost/responsibility center—an organizational unit which is given substantial autonomy for making its own business decisions but which is held accountable for its overall performance by management

feedback—information about the effects of system outputs which is fed back through the system as input thus notifying the system of required corrective action

informal organization—the "unofficial" ways in which people have been organized or have organized themselves to perform their work effectively

military-style organization chart—an illustration of formal lines of authority and rank within an organization

open system—a system which interacts with its environment by receiving inputs and producing outputs

self-correction—a characteristic of a system which enables it to adjust to new conditions in the environment which upset its self-regulatory process

self-regulation—a characteristic of a system which enables the system to maintain itself in a steady state

span of control—a method of organizational analysis which categorizes people and offices in terms of the number of people who are direct subordinates

steering committee—a group of representatives from different business areas within an organization that reviews and acts upon all requests and proposals for systems assistance

subsystem—a system within a larger system

system—a series of interrelated elements that perform some activity, function, or operation

systems analysis—the process of studying the network of interactions within an organization and assisting in the development of new and improved methods for performing necessary work.

systems development life cycle—a comprehensive process beginning with the systems definition of a problem and including all the subsequent development and implementation activities through final review and evaluation of the new system

systems development team—a representative group of people (from top management, the systems department, and major units being affected by the new system) that is responsible for developing a complete system or a particular phase or component of a system

QUESTIONS AND PROBLEMS

1. Problem:

Granite Stone Company has in recent years had difficulty organizing an effective systems department. Most recently, its systems staff undertook a major production planning system. After one and a half years of turmoil and conflict, the project was scrapped when a number of key members of the systems staff left the company.

In their resignation letters, the systems personnel called management's attention to a lack of management support for the project and to the "traditional" and "conservative" attitudes of key personnel in production planning and manufacturing.

The various users organized themselves to counter these accusations in a joint memo to management. They referred in particular to the "heavy-handed" and "untrustworthy" tactics the systems department used so that it could report directly to the executive vice-president for corporate planning.

Granite Stone management has decided to restructure and restaff the systems department so that it will report directly to the executive vice-president for corporate planning.

a. What potential benefits and risks do you see in management's approach, given this particular situation?
b. What alternative approach would you suggest for Granite Stone Company? Why?

2. As the lead analyst assigned to develop a payroll system for Futura Enterprises, you have been asked to give a management-level presentation of the steps to be followed in developing a new system. Prepare the script you will use (including diagrams and illustrations) in describing the systems development life cycle.

3. You have been assigned as a consulting systems analyst to work with the systems department of National Radiator Corporation. Your objective is to review the systems department's relationship with management and nonmanagement personnel and ensure that proper standards exist within the systems department. What factors will you focus on during your evaluation?

4. Organizational Analysis Problem:
   Two district sales managers, Jones and Farley, report to Perez, the regional sales manager. Each has a staff of thirty salespersons. Jones has organized his staff into three sales area groups and selected three persons as sales area coordinators. Farley, on the other hand, uses one person as a staff assistant. All other salespersons in her district report directly to her.
   It is generally agreed within the organization that Jones runs an efficient operation, particularly because his district's sales have increased in each of the last five years. Although Jones is considered a fair manager, he is also considered too cautious at times and does not appear to have much influence on higher-management decision making.
   Farley is considered a relatively powerful figure who consistently gets her point across to management. Last year, when the sales in her district were down, she was able to convince management to alter the method of warehousing inventory—to her decided advantage and Jones' disadvantage.
   As a systems analyst, you will be assigned to develop a new order processing system. Answer the following questions:

   a. How might you explain the apparent power differences between Jones and Farley? (Use charts if necessary.)

   b. What concerns do you have about working with these two
      individuals? How will you approach working with them?
   c. What other organizational information will you be interested
      in?
   d. Who else will you want to talk to?

5. Name Label Corporation would like to develop an integrated
   financial system that would affect the following departments:
   Budget, Personnel, Accounting, Payroll, and Planning. The
   senior vice-president for finance has asked you to:

   a. Describe why the Team Model approach would be well suited
      to this project
   b. Diagram the team organization, illustrating user, systems,
      and management participation

6. Top management at Newvue Camera International has decided
   to change from old, outmoded batch processing systems to contemporary online, data-base-supported systems. Red Freder-
   icks, Manager for Systems and Programming, wants to please
   management but is concerned because neither he nor his staff
   has ever worked with the new technology that would be required.
   Can you suggest an approach that Red might propose to
   management?

MINICASE—CONTEXT INTERNATIONAL, LTD.

Context International is a major manufacturer of water treatment,
purification, and filtration systems. The firm has expanded rapidly
in recent years through the acquisition of smaller firms engaged in
the same business or in closely related businesses.

   Financial management and control at the corporate level have
become increasingly difficult with each new acquisition. Management believes that the difficulties result from each subsidiary company having its own financial and accounting system. There is a
wide range of data processing support among the subsidiaries, from
reasonably sophisticated to nonexistent.

   You have been assigned to help corporate management develop
an integrated financial information system that will operate on a
corporate-wide basis. Your first assignments are:

   1. To assist in developing a methodology for organizing the sys-
      tems development function at the corporate headquarters and
      at the subsidiaries

   2. To specify a standard approach to developing the new financial
      information system

   Describe the approach you would use in completing these assignments. Also, based on your readings in the introduction and
chapter 1 of this text, provide examples of the approach that you
feel would best suit the needs of Context International.

NOTES

1. The effective use of teams during systems development is an important
   area of study in its own right. For a more comprehensive treatment of
   the topic, see Philip C. Semprevivo, *Teams: in Information Systems
   Development* (New York: Yourdon Press, 1980).

2

Effective Communications
in Systems Analysis

CHAPTER OBJECTIVES

When you finish this chapter you will be able to—

- *Identify and define the basic communication skills required in systems analysis.*
- *Discuss the importance of effective communication to the systems process.*
- *List the criteria for an effective question.*
- *Identify properly and improperly constructed questions.*
- *Discuss methods of questioning.*
- *Identify and define the three kinds of interview sessions.*
- *Define and discuss alternative types of interviews stating their advantages and disadvantages.*
- *Identify and discuss techniques for effective group communication.*
- *Identify and discuss factors needed for effective formal presentations.*
- *Discuss methods for improving communication skills.*

Today we seek technological solutions to major problems almost as a matter of course. The complex of human interaction which underlies these problems is frequently masked by charts and diagrams, and by promises of newer, faster, and more efficient machinery. Because of our failure to focus upon the human aspects of these problems, the promise of success through automation frequently remains just that—a promise.

In this chapter, we will discuss the skills and techniques needed for effective communication. This is an important step toward improved systems analysis. Effective communication can ensure that the right systems and procedures are developed and in a way which meets people's needs.

Certain kinds of background information about a systems problem can be obtained (and indeed are best obtained) through secondary sources, such as books, documents, and magazine articles or through questionnaires and memos. However, for the majority of detailed information required in properly assessing a problem and for assisting in its solution, there is no substitute for direct personal encounter with the people who live with the problem. There are a number of reasons for this:

1. It is difficult, if not impossible, to prejudge what is important to ask people if you are at a distance.

2. It is not always possible to obtain a concise written statement which is also complete.

3. The people who respond to written requests are seldom the ones doing the actual work.

4. It is difficult to follow up on written responses, which themselves might generate additional questions.

5. Working together implies a kind of mutual commitment to understanding which is difficult to generate in a letter or questionnaire.

Direct contact in the form of a dialogue or team meeting is a significant aspect of the systems analyst's work. However, the accuracy and reliability of those encounters will be improved if they are conducted with a high degree of skill in one-to-one and group communication.

Communicating throughout the Systems Development Life Cycle

Because the systems analyst is involved in the study of many different kinds and levels of functional interactions, he or she must be skilled in communicating with people at all levels in an organization. Since the analyst's work should not unnecessarily disrupt ongoing functions and responsibilities, the analyst must be both well organized and skilled in communications.

During the early stages of systems analysis when the analyst is attempting to collect and analyze data which will assist in problem definition, the analyst will spend a lot of time talking to the individuals who actually perform the job under investigation. Ability to learn about the job and understand the underlying problems will be directly related to the effectiveness with which the analyst is able to communicate. Factors such as gaining the trust of others, knowing how and when to ask meaningful questions about the job, and being able to listen effectively and observe will provide the analyst with accurate and reliable information. This accurate and reliable information is an absolute requirement for proper assessment of a problem. It enables the analyst to devise "good" plans which will assist in dealing with the problem.

Once the basic data has been collected and analyzed, the communications role of the analyst changes considerably. The emphasis is on a more directed and less continuous encounter. For example, during the development of a systems proposal, feasibility testing, systems design, and computer programming, the analyst may be dealing with a limited number of team members, department heads, and data processing personnel. A good deal of the analyst's time during this period will be spent performing technical systems work, such as designing files, forms, reports, and procedures. Also the analyst might act as the leader at team meetings, coordinating the work of others and ensuring that all persons working on the project stay on schedule and that they are informed of what others are doing.

During the final stages of the systems process as it approaches implementation of the new system, the need for continuous, one-to-one and group communication is once again very evident. As people begin to learn how to operate the new system and as they begin to encounter difficulties due to system "bugs" and directions which are not precise enough, they will require the immediate presence and help of the analyst. By being accessible and skilled in communicating with others, the analyst can do much to increase the user's confidence in the new system. Through listening and observing, the analyst will learn what last minute changes and modifications are required if the system is to function properly.

AN APPROACH TO EFFECTIVE COMMUNICATIONS

Initially the systems analyst's approach to communicating with another person should be governed by the desire to establish rapport with that person. Stated simply, rapport is a feeling of confidence and understanding between two individuals. It is the basis for effective communication.

We must be careful not to confuse rapport with pleasantness. An exchange of artificial niceties is little help in arriving at true understanding, particularly since real problems are often unpleasant to discuss. To always agree is not necessarily communication. Also, true communication is a two-way process, not just letting someone else talk.

To establish rapport requires sincerity and honesty, even when the topic is unpleasant. Sincerity and honesty, coupled with skills in listening, watching, and recording what is being said will enable the systems analyst to establish rapport with others.

Listening Skill

To critically and effectively listen to another person is a skill which is acquired only with considerable effort and some practice. Frequently, the inability to listen is the primary barrier to effective communication.

Recognizing the natural inclination not to listen is neither new nor startling. It has been observed historically as a source of both humor and malice. The social game "pass the word" is an example of the distortions which normally occur when a word or simple phrase is relayed from one person to another. Skill in listening can eliminate similar distortions and misunderstandings from systems analysis.

There are several well-known techniques you can employ to make listening truly effective:

- Make the other person central in the discussion by attempting to place yourself in his or her position.

- Give full attention and interest to what is being said.

- Organize what is being said by recognizing and concentrating on the *major* points which the other person is attempting to make.

- Listen to the implicit, as well as the explicit, meaning of what is being said. (Is the other person trying to tell you something which goes deeper than what is being said?)

- Minimize or shut out distractions by concentrating on the other person.

- Examine and discount your own prejudices.

- Show interest through questions and comments which encourage the other person to express ideas or to elaborate upon some earlier comment.

The Attentive Eye

Watching, like listening, is crucial to effective communication. By studying the physical actions and reactions which accompany verbal expression, you can often detect important feelings and attitudes which are not, for one reason or another, being voiced. It is wise to give these meaningful gestures equal weight with their verbal counterparts. In this sense, it is a good idea to look for and recognize:

- Excessive animation in the other person's responses
- Nervous tappings and laughs
- Decided frowns

- Undirected or widened eyes
- Nervous tics

When any of these normal responses are evident, proceed more slowly and cautiously. These kinds of responses may mean that the analyst's mere presence is being interpreted as a personal threat. The person being interviewed may need time to relax and regain his or her composure. In any event, it is better to over-evaluate a gesture's meaning and lose time than to under-evaluate it and lose communication.

Comments on Objectivity

It is not enough to simply use good logic. Even the most rigorous logic, when it is applied to data whose origins are biased and prejudicial, does little more than mask erroneous and unfounded thought. To deal effectively with this highly personal and complex matter, the systems analyst must strive to:

- Recognize that subjective personal viewpoints exist in all people.

- Determine the extent to which the person being dealt with harbors strong personal beliefs and ideas about others.

- Attempt to eliminate the negative effects of these subjective beliefs and ideas from the analysis.

Let us imagine for a moment that Ms. Harris is a capable and forthright marketing manager. Since she joined Sour Candy Corp. two years ago, she has made it a point to evaluate the interaction between marketing, sales, manufacturing, and production planning efforts of the company. The results of her evaluations have pointed up many operational deficiencies within the company and created a lot of ill will with other (mostly male) managers. As a systems analyst assigned to work in this area, you must eliminate from the analysis the data which is biased by the fact that Ms. Harris is a female, aggressive, and a critic. By not doing so you run the risk of building a system which is not based on fact.

In addition to dealing with prejudice in others, it is necessary to recognize and understand your own subjective view of the world. This is not an easy task. A certain amount of criticism and appraisal by others is especially important when your view is slanted by bias and prejudice. But there are some things that you can and should do for yourself. For example, consistent self-interrogation and self-criticism are helpful aides. Questions such as "Why do I like/dislike him?" and "Why do I always agree/disagree with her?" may be important tools for gaining objective self-awareness. A particularly interesting way to detect the existence of personal prejudice is to note the prejudices in others, jot them down, and then stop to look at them introspectively, as they apply to yourself.

The problems of gaining objectivity are many and complex. Certainly, an in-depth study is beyond the scope of this chapter, and yet its obvious relevance to effective communication warrants its mention. It is a problem which the systems analyst carries into each new work situation.

Documentation

There are two primary reasons why the analyst should take particular care in documenting what is communicated throughout the systems process. First, it is easy to forget both the exact content of what was said and the conditions under which it was said. Second, people remember some things better than others. This is frequently referred to as selective recall. By forgetting some things and remembering others that may have special meaning, the analyst can produce an incomplete and slanted view of what actually took place.

To avoid these pitfalls, the analyst should record fully the content of discussions with other people about the systems problem. This may be done by taking complete notes either manually or using a portable tape recorder. Of these two methods, recording equipment is more exact. However, it suffers from the dual disadvantages of being both awkward and at times personally offensive to the other person.

It is also vitally important to transcribe and organize recorded notes as soon as possible after leaving the other person. There are two reasons for this:

1. By reviewing the notes immediately, the analyst will remember much more of the encounter.
2. It is not always possible to record everything (particularly the atmosphere in which remarks are made).

The analyst should attempt to interpret all notes in a way that will preserve the implicit as well as the explicit meaning of what was said. Also, specific comments should be related to other important topics not discussed. For example, does the method for extending credit to customers, which was discussed, have implications for the processing of customer orders, which was not discussed?

Finally, all information should be summarized into clear, concise, and gramatically correct statements of what occurred. In summarizing, take care to place remarks in the order of relative importance to the problem-at-hand. Many analysts like to use such summaries as connecting devices between meetings with the same individual. The analyst presents the summary to the other person at the onset of the next encounter. In addition to its obvious benefits as a connector, this use of the summary frequently stimulates new comment from the other person. These new comments may cause the analyst to reassess the previously stated conclusion.

The Question Is All-Important

In the previous chapter we mentioned that the systems analyst is highly reliant upon others to answer questions about particular job functions (What? Why? Who? How?). Not all methods of questioning are equally effective for getting complete answers. In fact, some methods drastically reduce the possibility of obtaining a clear picture of the problem and therefore should not be used as the primary means of getting information. Some of these common errors in questioning are:

- Asking questions which can be answered yes or no. Reality is seldom this simple. The analyst who asks a yes or no question does not get a complete interpretation of how the other person views the problem.

- Asking unimaginative run-of-the-mill questions for which the other person has ready-made answers.

- Asking leading questions which suggest obvious answers. For example, "Don't you think that . . . ?" or "This could be used for that, couldn't it?"

- Asking non-neutral or emotionally charged questions which reveal the analyst's personal attitudes. (Sometimes voice inflections will "charge" questions, too!)

- Asking questions that are already answered or that serve no purpose except to waste time.

- Asking questions that are unrelated to the task at hand.

To avoid these pitfalls and to be truly effective, the analyst must ask questions that:

- Direct the other person to the right area.
- Allow the other person freedom in expression.
- Subtly prompt the other person to provide complete answers.
- Allow the analyst to remain neutral.

Questions may be both explicit (end with a question mark) or implicit (suggest that more information is needed without so stating). Five important methods which meet the above stated basic requirements for effectiveness include:

- Rephrasing, or paraphrasing as it is sometimes called (explicit)
- Indication of understanding (implicit)
- Encouragement (explicit *or* implicit)
- The pause or silence (implicit)
- Summarization (explicit)

Paraphrasing is, as its name implies, a thoughtful rearrangement of the other person's words:

Other: "This new method of staffing is the best thing that ever happened to the company."

Analyst: "How do you think the new staffing method will help the company?"

The purpose of this method is twofold. First, the analyst, by simply tossing back the other's statements, gives that person an opportunity to view them more objectively. In the previous example, was the new staffing method really that good for the company or were there other reasons for liking it? In any event, the question was open-ended enough to accomplish its second goal, namely, to invite some additional justification or qualification of what was said.

The indication of understanding is a method which attempts to assure the other person that he or she is being read properly. Consequently, the other person feels more confident about proceeding. It is a question in that it implies, "Yes, I understand you. Will you tell me more about it?" One potential difficulty in applying this method is that the analyst must be careful not to agree or disagree but to remain neutral in the response. Some sample responses which preserve neutrality include:

"I see."
"Yes, I understand."
"Uh huh."

It is acceptable to accompany these verbal responses with a neutral gesture such as a nod.

Periodically, it is important to use a more open kind of encouragement as a means for prompting the other person to continue or expand upon some earlier point. Phrases such as "That's interesting" or questions such as "How do you feel about this?" ask the other person to go into greater detail and to become a bit more personal about it.

In some ways, silence is the most difficult and yet the most effective means of questioning. The pause shows the other person that prior questions have not been completely answered. However, silence is extremely difficult to handle, and it will probably take the analyst some practice to feel at ease in using it.

It is the systems analyst's responsibility to periodically bring order to what is being said and to clarify the relationships between seemingly discrete facts. Since this should be done without interjecting too much of a personal tone, the analyst's summary should be more on the order of a reflection than an affirmation. For example, questions that summarize and organize can begin with:

"You feel that . . .?" or

"You would like to . . .?"

The systems analyst needs some level of sophistication in each of the above methods of questioning. Then the analyst will be able to conduct a more varied and a more interesting discussion or interview. Furthermore, he or she will be able to maintain a more reasonable pace of discussion and eliminate undue timelags.

Of at least equal importance to the manner in which a question is asked is the overall timing. A perfectly phrased question asked at an improper time is, at best, a wasted question and may defeat its own purpose. Consider the following points in timing a question:

- If a question seeks an answer which is already obvious to both parties, it is too late.
- If a question is not clearly understood by the other person, then chances are it is premature.
- When a question causes the other person to reflect upon his or her own words and to bring some logical order to what has been said, the timing is right.

In large measure, proper timing is learned only through experience. Seeing, too late, that the "ripe" moment has passed or feeling the wrath that comes with misunderstanding are good teachers providing one understands why these things occur and the ways to avoid them in the future. When communication problems arise, consider the bad timing of questions as a possible cause and improved timing as a potential correction.

USING THE INTERVIEW

The interview is an extension of the communications skills previously discussed and not a substitute for them. It is simply an attempt to structure one-to-one communications in a way which will better enable both parties to attain their mutual goals.

An interview may consist of one or more sessions (or meetings) which are usually fixed periods of time specifically set aside for uninterrupted discussion about a particular matter. Because of the complexity of problems in systems analysis, it is seldom possible to limit the interview to one or even two sessions. Rather, several sessions are often mandatory. Whenever possible, interviews should be organized to include:

- An introductory session
- Detailed session(s) (one or more)
- An exit session

The purpose and nature of each of these sessions is sufficiently unique to warrant some separate discussion.

The Introductory Session

The introductory session is particularly important in that it sets both the climate and the pace for all subsequent sessions. In effect, it provides a frame of reference. Failure here could retard progress and even jeopardize the success of the total analysis. During the introductory session, it is essential that the analyst accomplish at least the following:

- Impart an atmosphere of cooperation—a first step in establishing rapport.

- Outline the purpose of the interview and the course which the sessions will follow.

- Describe the general structure of the project and give some idea of the relative importance and placement of the interview to the total project.

- Outline what is expected from the other person (interviewee) and what can be expected from the analyst (interviewer).

- Remain sensitive to reactions from the interviewee. In the event that the reaction is negative, back off and take a little more time.

- Start to get into the underlying problem at hand, which is the purpose of the interview.

Whenever possible, all interview sessions should be kept to less than an hour. The initial session should probably be shorter than usual (say, less than 40 minutes) if possible. Interview sessions that are kept down in time tend to be more productive than lengthy sessions, since both parties remain alert and attentive throughout.

Detail Session

Just as the introductory session informs the interviewee, the detail session(s) provides the fundamental information that the analyst requires. The analyst will find it easier to get this information if he or she has developed communications skills similar to those discussed previously. Certainly, skill as an interviewer will be very important to the analyst's success in the detail session.

One final point worth mentioning concerns a "hang-up" experienced by most novice interviewers—namely, how do you start an interview? A point to remember is that it is best to begin in a direct, clear, and precise manner. Avoid undue fanfare at all costs. For example, once the hellos and handshaking have been completed, and both parties are seated, it is not unreasonable to give a brief pause and then to open with something like:

"(John), one of the things we should talk about today is the updating of the personnel master file; would you tell me how you add a new employee to the file?"

In so doing, we have quite naturally gotten to the core of what must be discussed, given the interviewee the floor, and at the same time made it easy for him to begin the discussion on the right track. This method of questioning always has a place in the interview and the beginning is no exception.

The Exit Session

In many ways the exit session is second in importance only to the introductory session. A special exit session is essential because:

- It officially brings together the many points covered in other sessions. Thus it brings these sessions to a formal close.

- It reinforces that accomplishments have been real and that they were possible largely through the cooperation between the analyst and the interviewee.

- It helps to avoid leaving the other person up in the air.

- It allows the systems analyst to outline the progress that has been made and the problems left to handle.

- It provides the other person with a summary and an opportunity to comment upon it.

An exit interview that accomplishes these tasks is rightly considered a vital segment in the interviewing process.

Handling Resistance

On occasion the analyst will encounter resistance from the interviewee. This resistance may be directed specifically at the analyst or to a general ideal which the analyst is seen to symbolize. This resistance may be mild, in which case it can usually be overcome, or it may be so severe that it breaks down the communications process. Mild resistance is generally caused by apprehension on the part of the other person because he or she fears that "sore points" will be touched upon by the analyst, or because the analyst's purpose is misunderstood. These occurrences can be overcome if the analyst is willing to back off a bit and give the interviewee a chance to realize that he or she has over-reacted to the problem. Other, more severe forms of resistance may emanate from such things as:

- Personality clash
- Conflict of social roles
- Power struggles
- Breakdown in communication
- Deep-rooted psychological problems within the interviewee

In at least the first four of the above, even a fairly strong resistance can be overcome by an interviewer who is honest in the quest to establish rapport. However, when resistance of this type is extreme, or when deep-rooted psychological problems are evident, it is probably best to seek alternate means for obtaining the required information.

One immediately available alternative is to admit the impasse and to recommend that another analyst be assigned to that portion of the project. A systems analyst, like anyone else, cannot expect to be equally effective in communicating with all types of people. It follows that there will always be a few with whom a particular analyst can find no common ground upon which to form a sound working relationship.

INTERVIEW TYPES AND TECHNIQUES

There are a variety of ways to conduct an interview. While some seem to be more useful in systems analysis than others, the analyst should have at least a basic understanding of the full spectrum of possibilities.

In one type of interview, the analyst has a prescribed set of standardized rules which include: the points to be covered, what to look for, what to report, suggested responses, and so on. This is called a *structured interview*. It is widely employed in many business situations, particularly in personnel departments, for example, with job interviews.

An opposite type is the *nonstructured interview*. The nonstructured interview is used with only a very general goal in mind and with few, if any, pre-established guidelines. Strong reliance is placed upon the other person to develop order and meaning throughout the interviewing process. Skill and subtlety in guiding that person become the primary contribution of the interviewer. Some forms of psychological and psychiatric counseling use the nonstructured type of interview.

In practice, there exist any number of interview types between the two extremes outlined above. And it is probably wise to view types of interviews as essentially falling somewhere on a continuum between highly structured and nonstructured. A more detailed comment on the potential benefits and deficits of these extreme types as they relate to systems analysis will be entertained shortly.

In addition to the type of interview employed, interviews can be categorized by the overall technique employed by the interviewer. Thus when little active comment and suggestion is provided by the interviewer, the technique is called nondirective. When the interviewer is more prone toward interjecting expert remarks and suggestions, the technique tends towards the directive.

Again, the practical analogy may be drawn to a continuum of alternative techniques ranging from the directive to the nondirective. Both extremes offer opportunities for successful communications in systems analysis and hence will be considered at a later point in greater detail.

To further distinguish between "types" of interviews and interviewing "techniques," it may be helpful to remember that the interview is a communications interaction between two persons. Types of interviews determine, through structuring, what will or will not be communicated. Techniques, on the other hand, determine the manner in which communicating will occur.

Table 2-1   Comparison of structured and nonstructured interviews

Type	Advantages	Disadvantages
Structured	1. Eases training new personnel	1. Requires high initial cost of developing quality standards
	2. Is more economical because sessions are shorter due to limited and defined scope	2. May dull interviewing spontaneity
	3. Is easier to evaluate and compare effectiveness and progress due to standardization	3. Relies on universal standards, which may not apply in most cases.
	4. May be conducted with some sophistication with limited training	
Nonstructured	1. Allows for greater spontaneity and creativity	1. Is more time-consuming; hence more expensive
	2. Provides greater insight into how the other person views the problem	2. Results in collection of more interesting information of little practical importance
	3. Allows for greater flexibility	3. Requires high level of individual training, experience and ability
		4. Can be unproductive in the hands of a novice

Structured versus Nonstructured Interviews

The chart in table 2-1 shows the relative advantages and disadvantages for both structured and nonstructured types of interviews. Interviewing is only one of a number of important skills necessary to effective communications and effective systems analysis. Because of the high skill requirement associated with nonstructured interviewing and because this type of interview results in the collection of considerable data which is of little systems value, it is recommended that the analyst use a more structured approach.

A sample of the standardized guidelines that might be used to structure interviews during the data collection phase of systems analysis follows:

Interview each member of the user department and focus on gaining information about each of the following:

1. *Office Organization*   How does the management organize its personnel? How does this organization relate to the major functions which the office performs?
   a)   Construct a general description of the major functions performed by the office.
   b)   Construct a chart which depicts lines of authority and responsibility and/or other kinds of formal and informal organizational interaction.

2. *Functional Flow*   For each important function determine the steps required and describe their significance.
   a)   Construct a procedure and/or flow chart which illustrates in detail the work that is performed.
   b)   Explain which personnel are involved in or affect the problems, decisions, and materials used at each step in the process.

3. *Resource Requirements*   Determine what resources are applied by the organization in getting the job done.
   a)   What personnel requirements such as specialized training and experience are required in doing the job?
   b)   What equipment and materials are required to support the work efforts of personnel?
   c)   How do the resource requirments utilized translate into a cost of doing business?

4. *Time Relationships*   How does the work performed relate to specific times of the year or other business cycles?
   a)   Are there workload peaks and valleys?
   b)   What are the actual work volumes by time?

5. *Forms, Procedures, and Reports*   What forms, written procedures, and reports are utilized in the course of performing work?
   a)   Include examples for each form, procedure, and report used.
   b)   Note whether the material is originated by the office, modified by the office, and/or transmitted to another office.
   c)   Make comparisons which determine usefulness, duplication and incompleteness.

6. *Nonexistent Desirable Functions*   Record the opinions of people as they relate to job improvement.

To answer all of the questions raised in the above guidelines, the analyst will necessarily talk to many people and may ask each of them many of the same questions. The purpose of the guidelines is to give direction to these discussions. Through structuring, the analyst is more likely to ask the right questions and less apt to overlook important areas of concern or collect irrelevant data.

Although the previous example represented a structured approach, it was by no means the most structured approach conceivable.

Directive versus Nondirective Technique

It is important not to confuse the term nondirective with undirected or nonparticipating. While active intervention and suggestions are at a minimum, involvement is at a maximum. The nondirective interviewer must remain mentally agile in order to get, through the use of a few well-stated and well-timed questions, complete and accurate information.

It is equally important to distinguish between directiveness and lecturing or closed-mindedness. With the directive technique, the analyst uses greater active participation, not domination.

Neither the directive nor the nondirective technique assumes that the interviewer knows the facts. To the contrary, acquiring and sharing knowledge is the purpose of the interview. Some of the major advantages and disadvantages of the directive and nondirective techniques are outlined in table 2-2. Since there are real advantages and disadvantages in applying either technique, it is helpful to base the decision to use a particular technique on what fits the specific systems needs and requirements.

Systems analysis is, in a sense, a process of education that depends upon effective communications. Because of this, it makes sense to be nondirective during the early interview sessions. In so doing, the analyst maximizes the opportunity for gaining complete information on a topic about which he or she knows little. Also, more listening than talking on the part of the analyst during the early stages helps to place the other person at ease and hence increases the possibility for quickly establishing rapport. In later sessions, the analyst can become more directive without threatening the communications process.

The real point to be made here is that the systems analyst should know and be able to use all the different interviewing approaches. This also means employing the right technique at the right time.

There is one additional and most important factor to be considered in the selection of an interviewing technique. To be effective,

Table 2-2   Comparison of nondirective and directive interview techniques

Technique	Advantages	Disadvantages
Nondirective	1. Enables the analyst to gain a better understanding of how the other person views the problem	1. Is more time consuming
	2. Provides greater detail on the problem	2. Requires more interviewing skill to be effective
	3. Helps to establish rapport; hence, reduces resistance	3. Often provides more information than is needed
	4. Provides better understanding of secondary factors, which may be important	
Directive	1. Makes interview more direct and to the point, hence shorter	1. Increases possibility that significant information will be overlooked
	2. Eliminates many points of limited importance	2. Provides less real understanding of the other person's thoughts and feelings
	3. Is easier for most systems analysts to work with effectively	3. Increases possibility of resistance from the other person

the systems analyst should be able to use the technique naturally. If the analyst is not at ease, he or she will find it difficult to establish rapport, and whatever communication may have existed will soon be eroded.

In trying any interviewing technique, expect a certain amount of awkwardness at first. With more knowledge and experience, the systems analyst will feel more confident in the interview situation. However, if a feeling of awkwardness persists for an extended period of time, it is a good idea to experiment by varying the technique.

EFFECTIVE GROUP COMMUNICATION

The widespread use of the team approach to systems development has added a new dimension in which the analyst must perform effectively. Left to themselves, team or group meetings frequently degenerate into gripe sessions or disorder. The systems analyst can play an important role in making these meetings orderly and productive, provided that he or she has acquired and uses the necessary knowledge and skills.

Techniques of primary importance in achieving effective group communication include:

- Conducting team meetings
- Generating creative ideas
- Achieving group decisions

Each of these topics will be discussed more fully in the following sections, because they have become increasingly crucial to effective systems development.

Conducting Team Meetings

Conducting a team meeting is best not left to chance. Rather, it is wise to structure the meeting in a way that gives it direction, ensures that all team members participate, and brings periodic closure by demonstrating that progress is being made.

It has been suggested[1] that team meetings can be made more effective by adhering to the following guidelines:

1. When possible, hold the team meetings early in the workday, when people are still "fresh." Also, reserve a certain time and day as the time for the team to meet. Meeting in the same place is preferred unless there is a specific reason to meet elsewhere.
2. Do not schedule the meeting unless everyone can attend.
3. Do not start the meeting until everyone is present.
4. Cancel the meeting if someone is still missing after the first twenty minutes.
5. Prepare a detailed agenda (and preferably distribute it) before each meeting.
6. For each topic, actively solicit criticism and comment from *all* team members.
7. Encourage the team to produce tangible products (however tentative) instead of just talking about what ought to be done.

8. Limit the meeting to an agreed-on length.
9. Insist on completing one agenda item before going on to the next.
10. Be sure that each person leaves the meeting with something to do before the next meeting—either individually or with others.
11. Have one person serve as the meeting recorder and publish daily minutes that highlight agreed-on items.

Two other techniques can be particularly helpful in conducting the meeting.[2] First, use large sheets of paper or chart pads to display anything significant as it occurs throughout the meeting. This technique helps people grasp what is really going on, helps maintain discipline, imparts a sense of direction, prevents important issues from being forgotten, and provides a useful form of documentation.

Second, dividing the team into subgroups is useful for collecting data, working on specific items that do not require the full team, and making progress when the full-team approach bogs down because people cannot agree.

The above methods and suggestions can help make team meetings productive and worthwhile.

Generating Creative Ideas

One of the most obvious reasons for wanting to use a team approach is that it creates an opportunity for many people with very diverse backgrounds to participate in systems development. Thus, if the interaction among these individuals is properly orchestrated, it should generate a greater number of creative and qualitatively superior ideas.[3]

The systems analyst can use several techniques to help the group maximize its potential for generating high-quality, creative ideas. Two of these techniques, namely, brainstorming and the nominal group process, are described below.

In a brainstorming session, one team member acts as a reporter. Standing at the front of the room, in full view of the team, the reporter writes on a chalkboard or large pad the other team members' ideas exactly as they are stated. A very important rule of brainstorming is that no evaluation of the ideas being expressed is allowed. If someone disagrees with an idea or tries to change the wording of an idea, the recorder must remind the person that discussion will be permitted only after the brainstorming session. The high degree of interaction involved in brainstorming can produce many creative ideas. However, the technique will not work if less-vocal people fail to participate or if the recorder fails to keep people on track. Remember, brainstorming is a way of generating many new ideas and should be used only when that is the desired result. After the brainstorming session, open discussion about the ideas is important to clarify and weed out certain ideas.

The nominal group process is in many ways a silent version of brainstorming. Rather than asking people to state their ideas aloud, the team leader asks each member to jot down a minimum number of ideas—say, five or six. In the second stage of the process, a single list of ideas is composed, usually on a chalkboard or chart pad. This list is placed in full view of the group, and the group discusses it, as in the period after a brainstorming session.

The nominal group process is easier to control than brainstorming and has the added benefit of eliciting ideas from team members who are shy or reluctant to speak up in group situations.

Achieving Group Decisions

Frequently, team members individually decide where they stand on an issue before there has been much group discussion. If many team members do this and their decisions conflict, then it may be difficult to achieve agreement. For this reason, it is important to look at group decision making as being more than just the act of deciding.

The systems analyst needs to encourage the team to look at decision making as a three-step process that includes:

1. Data collection—the team should take time to collect all the relevant information pertaining to the issue. Team members who came to the meeting with a preference should be allowed to present their ideas and opinions to the team.

2. Data aggregation and evaluation—the data should be summarized, compared, and evaluated. Techniques for quantifying the data, such as ranking their importance (for example, first, second, and so on) or rating them (for example, on a scale of 1 to 10), should be encouraged instead of less precise verbal evaluations.

3. The act of deciding—there are a number of different group decision-making methods, but not all are equally suited to systems development decisions. In the order of their preferred use, these methods are:[4]

   Unanimous consensus
   Consensus
   Majority rule

Minority rule (subgroup decision)
Authority rule
Lack of response (letting the issue die)

It is recommended that important decisions be made using majority rule as the minimally acceptable basis. Preferably, a consensus should be reached, because it reflects a broad base of support for the resulting decision. The wise team leader frequently will not accept as legitimate any decision that lacks this consensus.

EFFECTIVE PRESENTATIONS

The systems analyst is frequently involved in making formal presentations to management, users, or technical personnel. The time allotted to convey the intended message of these presentations is frequently short. To be well received, the presentation must be:

- Restricted to only those topics of major importance and concern
- Delivered in a direct and pertinent manner that is geared to the level of the audience
- Expressed in terms that the audience readily understands
- Presented in a cohesive, logical, and aesthetically pleasing sequence

Also, it is a fact that well-conceived charts and graphs help to transmit more information in less time. These visual aids are more aesthetically pleasing than lists of words and more entertaining than simply listening to someone else talk. In making formal presentations, it is highly recommended that the analyst use professionally prepared charts and graphs. If a trained artist is not available, someone with artistic talent should be asked to assist in making these aids. The aids may be placed on a flip chart, transparencies, or (if time and money permit) photographic slides. In preparing visual aids, these guidelines should be followed:

1. Use large print.
2. Be concise—avoid verbiage.
3. Use color or underlining to highlight major points.

There are several types of effective charts and graphs, each suited to a particular use. Examples of these follow.

The *pie chart* (as shown in figure 2-1) is particularly effective in illustrating how resources are being used or what contributions are being made and by whom.

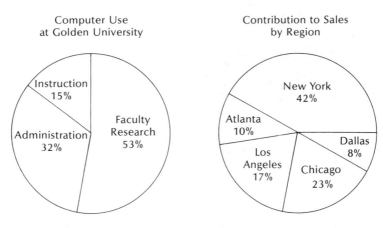

Figure 2-1 Examples of pie charts

The *bar graph* is particularly useful in presenting scaled comparisons over time; figure 2-2 presents an example.

The *line graph* is an easy-to-construct and effective way to present trends. With two or more lines, it may show comparisons in much the same way that the bar graph does. However, other than its ease of construction, the line graph offers no advantages in showing comparisons and has some disadvantages—for example, it can be difficult to follow if the lines begin to cross. The graph illustrated in figure 2-3 uses a single line.

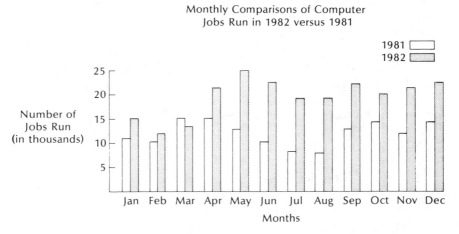

Figure 2-2 Example of a bar graph

Figure 2-3 Example of a line graph

Pinwheel diagrams are very effective in illustrating the subsystem components constituting a larger system. At the outset of a presentation, they can also tie together all the points that will be covered, or at the end, they can summarize all the points that were covered. Figure 2-4 presents an example.

Figure 2-4 Example of a pinwheel diagram

Besides the techniques discussed above, note that the standard charting techniques used in systems analysis, such as flowcharts and HIPO diagrams (both of which are covered in chapter 3), can also be used in formal presentations.

Making effective presentations is well within the capabilities of most people. The keys are careful planning using the list of suggestions presented earlier in this section, professional-looking visual aids, and practice. This final point about practice needs to be emphasized. Before being delivered, a presentation on a topic as complex as a new system deserves several practice runs. Do not attempt to "wing it." Also, be sure to leave sufficient time to correct the visual aids in case flaws are discovered in them during the practice runs.

IMPROVING YOUR COMMUNICATION SKILLS

The improvement of your ability to effectively communicate with others is your responsibility. Without your sincere interest, even a comprehensive training program, such as that discussed in appendix A, is of little consequence. Given self-motivation to communicate with others more effectively, you can make the following ingredients work for you:

- Research
- Practice
- Self-awareness
- External criticism

There is not a great deal of existing literature which deals specifically with the topic of effective communications in systems analysis. However, there are a number of important works which have been developed regarding effective communications—both in general and as it relates to specific professional fields, such as business management and counseling. By reading what is available, you will gain a greater understanding of how to communicate effectively.

To develop skill in listening, questioning, watching, and interviewing, you must practice what you have learned. For example, plan to practice your phrasing of questions during your casual conversations and in the classroom. Then, sharpen your listening skills by making a conscious effort to concentrate upon and organize what is being said. Also, it is possible to contrive mock interview situations with the assistance of two friends. One of you can assume the role of the systems analyst while the second party assumes the role of a user, such as the college registrar. The user is then interviewed regarding a particular problem area (for example, the registration of students). A silent third student evaluates the actions and reactions of both the student analyst and the user. At the conclusion of the mock interview the student evaluator presents his or her critical impression of the interview.

The third important ingredient for improving your communication skills is self-awareness. You must strive to critically evaluate your *self*. By knowing who you are, what you stand for, and what your hang ups are, you are better able to understand what your personal impact will be on other people and, perhaps, why it is so.

External criticism of your actions by others can also have a positive influence upon the development of your communication skills—particularly when the critic has already acquired the necessary skills and knowledge. Recognizing beforehand that you need to improve these skills will help to ease the sting of external criticism. Expect to receive criticism and attempt to benefit from it.

Certainly the development of effective communication skills is not an overnight process. It requires real determination and a positive attitude towards other people. This chapter has pointed out several important reasons for seeking improved communications in systems analysis. The ability to communicate effectively is well within the capabilities of most people, providing they give attention to developing the necessary skills. There is nothing mysterious about the communications process.

CHAPTER SUMMARY

It is necessary for the systems analyst to acquire ability in communicating with people at all levels in an organization. By developing basic communication skills in listening, watching, documenting, and questioning, the analyst will be able to obtain accurate and reliable information about the problem and possible solutions to the problem. This will provide a credible basis for developing an effective new system.

There are a variety of interviewing types and techniques which can help the analyst to structure the process of communication. However, the burden for gaining effectiveness in the use of these and other communications skills rests with the individual. That is, improving one's ability to communicate effectively requires as a prerequisite some self-motivation as well as a systematic program for improvement.

WORD LIST

brainstorming—an oral technique used by teams to generate many new, creative ideas

decision making—a three-step process involving data collection, data aggregation and evaluation, and the act of deciding

detail sessions—those interview sessions which provide the fundamental information that the analyst requires in developing a new system

directive—an interviewing technique in which the interviewer is more prone towards interjecting expert-based remarks and suggestions

exit session—the final interview session which brings an official and proper close to the interview

interview—a formal structuring of one-to-one communications towards some mutual goal consisting of a number of sessions

introductory session—the first in a series of interview sessions which sets both the climate and the pace for all subsequent interview sessions

introspective—to view the behavior and ideas of others as they apply to oneself

nominal group process—a mainly written technique used by teams to generate new, creative ideas; involves soliciting written contributions from all team members and then discussing the contributions

nondirective—an interviewing technique in which the interviewer does not provide active comment and suggestion

nonstructured interview—a type of interview in which only a very general goal is specified and with few, if any, pre-established guidelines

paraphrasing—a method of questioning which is a thoughtful rearrangement of the other person's words

rapport—a feeling of confidence and understanding which exists between two people

structured interview—a type of interview in which the analyst is provided with a prescribed set of standardized rules for conducting the interview

QUESTIONS AND PROBLEMS

1. Problem:
   Sissons is a systems analyst who has just joined the company. In his first assignment, you have noted the following about his approach:

   a. He tends to avoid or limit encounters with user personnel whenever possible.
   b. He appears quite uncomfortable in a one-to-one encounter, which frequently creates difficulties for the person he is interviewing.
   c. He tends to overlook crucial issues or treat them lightly.

   If you were a senior systems analyst asked to help Sissons, what would you recommend?

2. Joe Perkins, Systems and Programming Manager for Nusound Radios Corporation, has determined that his newer employees would benefit from hearing some of the more experienced systems analysts talk about effective interviewing. Joe has asked you, as one of his senior analysts, to prepare a presentation covering the following topics:

   • Effective listening
   • Effective questioning
   • Directive and nondirective interviewing techniques
   • Using the structured interview
   • Types of interview sessions

   Outline the points you will make for each of the above topics. Include examples where appropriate.

3. You have been asked to work with managers in your firm's Manufacturing Division to help generate a number of ideas for new data processing applications to the manufacturing process. What types of group process techniques might you use? How would they work?

4. James Soote Air Filters, Inc. has been using a team approach to develop a marketing system. Fred Weatherby, from the advertising department, is a compulsive talker and tends to dominate discussions of new and improved ways of doing work. What group process technique would you use in these discussions to elicit other people's ideas? In the longer term, how can

you make group meetings more effective or do without them altogether?

5. Evans Pineapple Company's use of data processing has increased by these percentages in the last five years: year #1, 26%; year #2, 37%; year #3, 18%; year #4, 22%; year #5, 42%. The systems department believes that this trend warrants major expansion of the firm's computer equipment. You have been asked to assist in a presentation to management. Your specific tasks are:

    • To prepare a chart or diagram illustrating the firm's increased use of data processing
    • To develop a brief script for a two-minute presentation

    The major data processing users are as follows:

Accounting	37%
Personnel	18%
Manufacturing	5%
Marketing	20%
Research and Development	12%
Others	8%

    How would you present this information?

6. Ned Newcomer, a systems analyst in your department, has been complaining about the effectiveness of the user team members working with him on his project. It sounds as though little progress is being made, and you suspect that the problem may be Ned's approach and not the team members (several of whom you know and respect). You would like to help Ned, so you construct a series of questions about team effectiveness issues to ask him. What questions would you ask about:

    • Effective team meetings
    • Generating creative ideas
    • Achieving group decisions

    Be prepared to interview Ned on these issues.

7. You have been asked to interview the controller at Spicy Soups, Inc., who is responsible for the payroll, accounts payable, accounts receivable, purchasing, and general accounting functions. You know very little about this area of the business and will need to rely heavily on the controller for information and guidance.

a. Develop objectives for your first interview with the controller. Determine quite specifically what information you would like to obtain from him.

b. Develop one or more questions for each objective. Remember the requirements for proper questioning that were presented in chapter 2.

8. Discuss each of the following aspects of achieving effective group communications:

a. Structuring and conducting group meetings
b. Generating creative ideas
c. Achieving group decisions

9. Discuss ways of improving the overall effectiveness of formal presentations. Provide examples of various visual aids that the presentor might use effectively.

MINICASE—CONTEXT INTERNATIONAL, LTD.

As a part of your continuing assignment with the Context International corporate staff, you have been asked to prepare a brief (5–7 minute) presentation to the chief financial officers (CFOs) of each of the firm's eight subsidiaries. The topic of your presentation is "developing an integrated financial system."

Develop your presentation—including appropriate visual aids—around the following points:

• The firm's growth—sales have grown from $50 million to $250 million in seven years; sales are projected to increase by 22% per year over the next three years.
• The lack of automation—the subsidiaries are 0–60% computerized in the following areas:
   Payroll
   Accounts payable
   Accounts receivable
   Cost accounting
   General accounting (for example, general ledger, corporate consolidations, and so on)

- An overview of the functions to be included in the new integrated system
- The methodology to be used during systems development and implementation
- The effect of new systems implementation on computer resource requirements

NOTES

1. Philip C. Semprevivo, "A Critical Assessment of Team Approaches to Systems Development," *Proceedings of the Fifteenth Annual Computer Personnel Research Conference* (New York: Association for Computing Machinery, 1977), pp. 94–97.
2. For these and other helpful suggestions, see J. K. Fordyce and R. Weil, *Managing with People* (Reading, Mass.: Addison-Wesley, 1971), pp. 157–67.
3. For a comprehensive summary of research comparing individual and group performance in problem-solving situations, see M. E. Shaw, *Group Dynamics* (New York: McGraw-Hill, 1976), pp. 51–81.
4. E. Shein, *Process Consultation* (Reading, Mass.: Addison-Wesley, 1969), p. 13.

3

Tools of the Analyst

CHAPTER OBJECTIVES

When you finish this chapter you will be able to—

- *Describe the structured, top-down approach to systems development and its advantages.*
- *Construct a VTOC diagram and IPO diagrams, given a system description.*
- *Provide a functional description of a process, given a data flow diagram.*
- *Provide a functional description of a process, given a Nassi-Shneiderman chart.*
- *Identify and define the ANSI standard flowchart symbols.*
- *Provide a functional description of a process, given a systems flowchart for that process.*
- *Construct a systems flowchart, given a functional description for a process.*
- *Construct a decision table using the quadrant method, given a functional description for a problem.*
- *Discuss the various kinds of decision tables and their uses.*
- *List the important structural considerations in developing a fixed-format questionnaire.*
- *Construct and issue a fixed-format questionnaire and prepare a report on the survey results.*

In the previous chapter we discussed several important communication skills which should be mastered by the systems analyst. In addition, the analyst must learn how to use effectively a number of other systems tools. This chapter will focus on both structured and traditional systems tools which have been basic to effective work within the profession.

Several of these tools are widely used in business, government, and other social institutions for both technical and nontechnical purposes. The current discussion is primarily concerned with their application to systems analysis.

STRUCTURED, TOP-DOWN APPROACHES

One of the most significant advances in systems analysis during the last decade has been the introduction and growing use of so-called

structured, top-down approaches to analyzing and designing information systems. Although these approaches were initially introduced as computer program design and coding aids, they have
increasingly been adapted for use in systems analysis and design.

There are broad differences in the symbols employed in various
structured charting techniques, yet they share a common set of
beliefs or ideology:

- It is important to identify a system's structure before attempting
  to define or describe the system's processes.
- One should first describe the system's structure in a very general
  or overview fashion and then break its structural components into
  smaller, more detailed components (hence the name top-down).
- Upon completing a structural taxonomy of the system, one should
  describe the system's processes, again employing a top-down
  approach.

In general, the use of structured, top-down approaches has
improved the orderliness and quality of systems development. These
approaches have also improved the quality of documentation and
thus facilitated communications among project team members and
communications with others. Finally, these approaches have been
an effective means for dividing or modularizing very large systems
into smaller, more manageable subsystems without sacrificing future
opportunities for integrating subsystem components into a single,
comprehensive entity.

Several new charting techniques have been developed specifically
to support the goals of the structured approach:

- The HIPO technique
- Data flow diagrams
- Nassi-Shneiderman charts

In the future, one can expect greater use of these and other structured charting techniques. One can also expect continued use of the
more traditional charts and diagrams, such as flowcharts and decision tables, because they can be effective in many instances. However, these traditional techniques will be used increasingly within
the context of an overall structured approach and will not be the
primary techniques.

THE HIPO TECHNIQUE

The hierarchy plus input-process-output (HIPO) technique is a tool with broad uses throughout the systems development life cycle. It can be used in documenting existing systems, organizing new systems requirements, designing new systems and programs, and supporting training and implementation.[1]

The HIPO technique involves two different charting techniques: the visual table of contents (VTOC) diagram and the input-process-output (IPO) diagram. In applying the HIPO technique, one begins by constructing the VTOC diagram and then the IPO diagrams. For this reason, we have chosen to describe these techniques separately and in the order in which they would normally be used.

The VTOC Diagram

The VTOC diagram is used to develop a hierarchical structure through which systems functions and their interrelationships can be observed, discussed, and understood. Developing the VTOC diagram in a top-down fashion is the first objective in the HIPO technique.

Imagine for a moment that you are a systems analyst with overall responsibility for developing a comprehensive human resources information system. To use the HIPO technique effectively, you would, very early in the analysis, identify the major functions involved in the system, as shown in figure 3-1.

As mentioned earlier, an advantage of this approach is that it

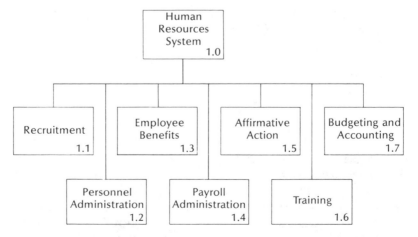

Figure 3-1 A VTOC diagram showing the major functions involved in a human resources system

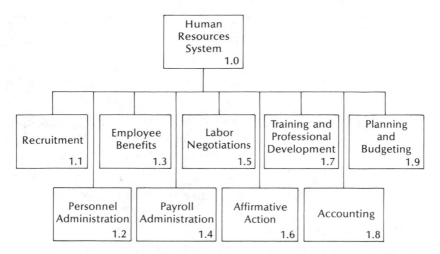

Figure 3-2 A revision of the VTOC diagram shown in figure 3-1

permits the analyst first to obtain a consensus about what the system's major functional components are. Through interaction with management and user personnel, the analyst can quickly restructure the VTOC diagram if necessary. For example, discussions with management may indicate that the major functions should be revised as shown in figure 3-2.

Two points about constructing a VTOC diagram are worth noting at this time. First, the numbering system denotes each function's level in the overall functional hierarchy. Thus, all the functions in figure 3-2 are major, first-level functions within the human resources system. Written in outline form, they would appear as follows:

1.0 Human resources system
    1.1 Recruitment
    1.2 Personnel administration
    1.3 Employee benefits
    1.4 Payroll administration
    1.5 Labor negotiations
    1.6 Affirmative action
    1.7 Training and professional development
    1.8 Accounting
    1.9 Planning and budgeting

Second, the VTOC diagram is most effective when a complete picture is rendered to fit on a single page or less. The VTOC diagram in figure 3-2, for example, displays the major components of the total system on less than a page.

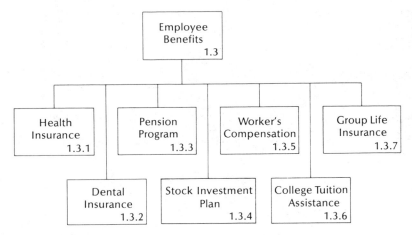

Figure 3-3   A breakdown of the employee benefits function shown in figure 3-2

Normally, in the next step of the HIPO technique, you (or others whom you assign) would functionally decompose these very global functions into a series of subfunctions. For example, the employee benefits function could be broken down into the various benefit programs, as shown in figure 3-3.

Discussions with management and user personnel may once again cause you to expand (or contract) the various subfunctions to be included in the employee benefits VTOC diagram. Or perhaps greater levels of detail must be specified. For example, a number of health insurance program options may be available to employees, each with its own set of unique requirements. One way of dealing with this situation would be to develop more detailed levels of VTOC diagrams, as shown in figure 3-4.

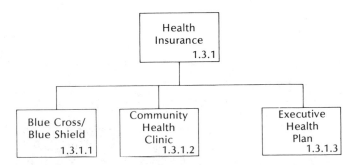

Figure 3-4   A further breakdown of the VTOC diagram shown in figure 3-3

In outline form, our revised VTOC diagram would appear as follows:

1.0 Human resources system
    1.1 Recruitment
    1.2 Personnel administration
    1.3 Employee benefits
        1.3.1 Health insurance
            1.3.1.1 Blue Cross/Blue Shield
            1.3.1.2 Community health clinic
            1.3.1.3 Executive health plan
        1.3.2 Dental insurance
        1.3.3 Pension program
        1.3.4 Stock investment plan
        1.3.5 Worker's compensation
        1.3.6 College tuition assistance
        1.3.7 Group life insurance
    1.4 Payroll administration
    1.5 Labor negotiations
    1.6 Affirmative action
    1.7 Training and professional development
    1.8 Accounting
    1.9 Planning and budgeting

The extent of functional decomposition is largely guided by the analyst's judgment of what is required at various phases in the systems development life cycle. Thus, during the early stages of data collection, the extent of detail required would be guided by the need to fully understand the business process; during program development, greater detail would be required in providing strict guidelines for coding various computer programs.

Quite apart from the rest of the HIPO package, the VTOC diagram can be of major utility to the analyst. This diagram has a number of major advantages. First, it is extremely easy to draw and thus can be used experimentally. It generally takes several tries to get the diagram into a form that is accepted as accurate, but each diagram is so easy to draw that changes can be readily accommodated. Second, the diagram lends itself to group construction and thus is a valuable aid when several people must participate as a team. Third, the diagram is also important in structuring and documenting that participation. Even when the "group" is composed of only the analyst and one other person (such as in the interview situation), the analyst may want to use the diagram to structure the interview. This approach helps the analyst avoid jumping from one topic to another at the whim of the interviewee and provides useful and concise documentation of the topics covered during the interview. Fourth, the diagram is potentially useful for making presentations at various levels of detail, depending on the intended audience. The diagram can clearly illustrate the hierarchical relationships among various functions and subfunctions. Finally, the VTOC diagram can provide a basis for continuous and evolving documentation and contribute greatly to overall project control. If developed properly during the early stages of analysis, the VTOC diagrams will be a valuable starting point for systems design and later for computer program design. Furthermore, the person with overall responsibility for a systems development project can use the structural overview in making job assignments and evaluating project progress.

IPO Diagrams

The VTOC diagram identifies what functions are being performed or need to be performed in a system. It also enables us to observe the hierarchical relationships among these functions. To answer other important systems questions, such as, "How is the function performed?," a second type of HIPO diagram is used, namely, the IPO diagram.

IPO diagrams describe functions contained in the VTOC diagram. As an identification or label, IPO diagrams carry the number of the VTOC box they represent.

IPO diagrams may present an overview or a detailed level. The recommended approach is to begin by constructing overview IPOs and then working down to greater levels of specificity. Also, one should generally begin by identifying the outputs of the function, then the inputs, and finally the processes. Figure 3-5 illustrates a complete IPO diagram.

An advantage of the IPO diagram is that it permits a more extended description of processes in ordinary English than a VTOC diagram does. Thus, the processing steps may be briefly described in the process box and more fully explained in the extended description notes at the bottom of the HIPO form, as shown in figure 3-5.

The symbols that describe inputs and outputs resemble standard flowcharting symbols. There are potential disadvantages in this reliance on verbal description and flowchart-like drawings; in particular, the documentation time required to describe all the functions may be very high. This drawback has led some analysts to employ a mix of diagramming techniques together with the VTOC

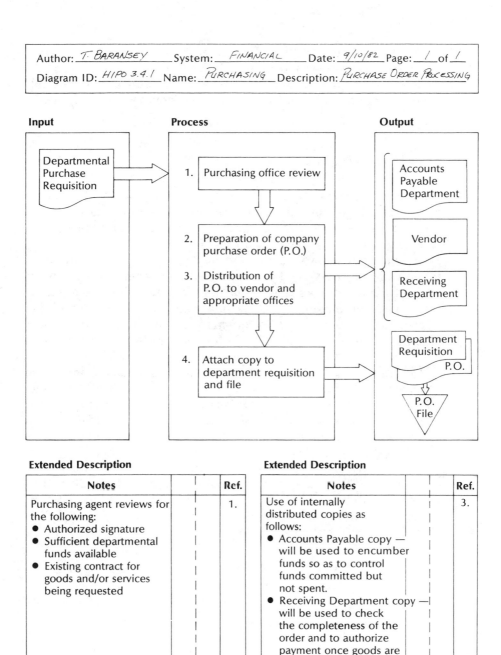

Author: _T. BARANSEY_____ System: _FINANCIAL_____ Date: _9/10/82_ Page: _1_ of _1_

Diagram ID: _HIPO 3.4.1_ Name: _PURCHASING_ Description: _PURCHASE ORDER PROCESSING_

Input	Process	Output

Input

Departmental Purchase Requisition

Process

1. Purchasing office review

2. Preparation of company purchase order (P.O.)

3. Distribution of P.O. to vendor and appropriate offices

4. Attach copy to department requisition and file

Output

Accounts Payable Department

Vendor

Receiving Department

Department Requisition

P.O.

P.O. File

Extended Description

Notes		Ref.
Purchasing agent reviews for the following: • Authorized signature • Sufficient departmental funds available • Existing contract for goods and/or services being requested		1.

Extended Description

Notes		Ref.
Use of internally distributed copies as follows: • Accounts Payable copy — will be used to encumber funds so as to control funds committed but not spent. • Receiving Department copy — will be used to check the completeness of the order and to authorize payment once goods are received.		3.

Figure 3-5 A complete IPO diagram

aspect of HIPO. These analysts use IPO diagramming only when its extended-description advantages are judged to be worth the added documentation effort.

As mentioned earlier, the HIPO technique is only one of a growing number of structured techniques. It represents a complete methodology with many advantages, including ease of use. In the following sections, alternative structured approaches are briefly presented.

DATA FLOW DIAGRAMS

Data flow diagrams are an increasingly popular alternative to IPO diagrams.[2] Data flow diagrams may be effectively used with a VTOC diagram to describe the process for a particular function or illustrate the data flow between functions.

Because they use very few symbols, data flow diagrams are easy to draw. Also, they lend themselves to describing logical flows without having to note whether or not the data will be on a particular computer medium, such as magnetic tape. This feature is important, because it enables the analyst to develop strong logical foundations for a system design before dealing with the more specific, physical issues.

A complete data flow diagram can be constructed using only three basic symbols, plus arrows that specify the direction of data flow, as illustrated in figure 3-6. As with other charting techniques, data flow diagrams can be developed at increasing levels of detail. In figure 3-6, it is apparent that the diagram is a more detailed description of a higher-level process whose reference number is 4.0 and which is entitled "Apply Payment to Outstanding Receivables." We can anticipate that, at some future point, it may be desirable to diagram in greater detail some of the processes shown on that diagram.

The data flow symbols are described individually below. You should become familiar with them.

```
┌─────────────┐
│             │
│ a.          │
│ Customers   │
│             │
└─────────────┘
```

The above symbol describes external sources and destinations of data. In a business system, the sources and destinations are frequently people, organizations, and other interacting systems. The symbol is given a meaningful name, such as "customer." For easy reference, the symbol may also be identified by an abbreviation, such as the lowercase letter in the upper left corner of figure 3-2.

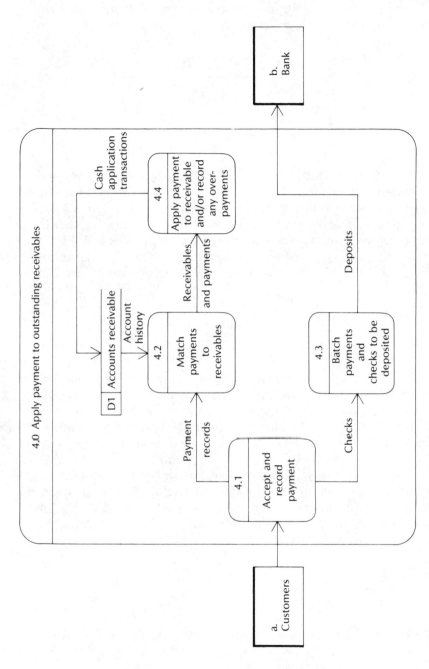

Figure 3-6   A complete data flow diagram

A source or destination may be shown several times, if necessary, to avoid creating crossed lines.

```
      4.0
     Apply
    Payment
      to
  Outstanding
  Receivables
```

   The above symbol describes, in ordinary terms, the process to be performed. This symbol is frequently identified by both a process name and number. The process number is simply assigned from left to right and from top to bottom across the chart. The number is not intended to relate to the overall processing sequence. As with any charting or diagramming technique, in a data flow diagram it is important to make the process name as descriptive as possible.

```
| D3 | Accounts Receivable File |
```

   The above symbol identifies places where data are being stored between processes. A readily understood name and an abbreviated identifier should be selected. (In this example, the designer decided to use a "D" as the first character of the two-character identifier for all data storage symbols.) A data storage location symbol may be shown several times, if necessary, to avoid creating crossed lines.
   Because data flow diagrams are not linear (straight-line), you must take care not to describe too much detail on a single diagram, or it may become overly difficult to follow. Beginning with an overview and working down to individual, more detailed levels of data flow is strongly recommended.

NASSI-SHNEIDERMAN CHARTS

Nassi-Shneiderman (N-S) charts (named after their codevelopers) have been primarily used for designing structured programs. Therefore, we will cover them only to the extent necessary to read the N-S charts constructed by computer programmers. We will primarily consider several of the unique symbols used in constructing N-S charts.

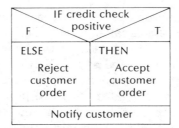

IF-THEN-ELSE or Decision Symbol

The above symbol neatly combines both the conditions and the actions associated with decision making. Thus it is almost immediately clear that, depending on the condition, we will either reject or accept the order, but, regardless of the condition, we will always notify the customer of the decision that is made.

DOWHILE Symbol

So long as a condition holds true, we may wish to perform a particular process (this is called a **DOWHILE** loop). In the example above, we will perform the credit check until there are no more customer orders to process.

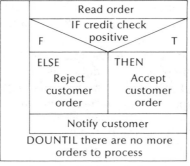

DOUNTIL Symbol

We may wish to perform a particular process until we encounter a particular condition (this is called a DOUNTIL loop). The example on the bottom of page 91 illustrates how the order credit check process could have been constructed as a DOUNTIL process.

Although N-S charts are not extensively used in all aspects of systems analysis today, expanded uses will probably be found for them in the future—particularly when the analyst is looking for a better way of clearly describing to others some of the more complex logical problems involved in analysis and design.

STANDARD FLOWCHARTS

Flowcharts provide the analyst with an effective means for:

1. Illustrating the logical order in which things happen for a complex process.
2. Visually pinpointing what resources and major decisions are required throughout the process.
3. Graphically explaining what could happen as a consequence of varying conditions and decision alternatives.

There are two major kinds of standard flowcharts which will be discussed in this chapter, namely: (1) the systems flowchart and (2) the program flowchart. In most instances the real difference between them is the level of detail being illustrated. In addition to standard systems and program flowcharts, a number of special-purpose flowcharts which are used in the collection and analysis of data will be discussed in chapter 5.

Systems flowcharts present a general overview of an entire system or process. Because they are more general, they can frequently be shorter than a detailed program flowchart. The systems analyst is able to use the systems flowchart in explaining a complex problem to nontechnical personnel or as a way of introducing technical personnel to a new problem topic. Figure 3-7 is an example of a systems flowchart.

Both kinds of flowcharts are systems tools. It is simply a question of determining which level of detail is required and for which audience. To construct a lengthy and detailed program flowchart for presentation to corporate top management would probably obscure rather than clarify the problem for them. On the other hand, a general systems flowchart, by itself, does not contain the level of detail required for computer programming.

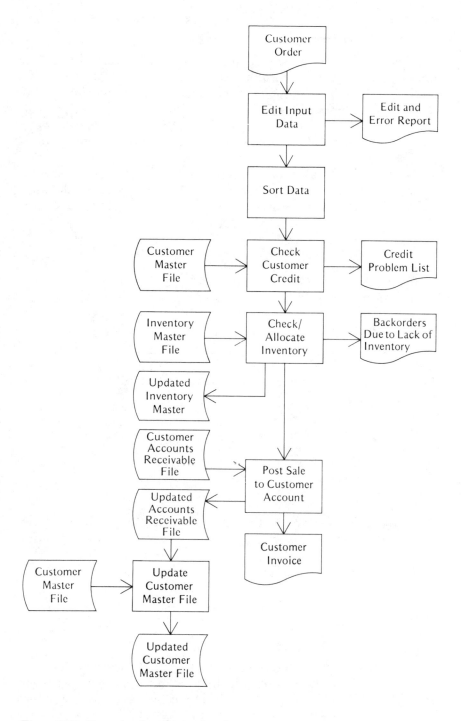

Figure 3-7   Example of a systems flowchart

Flowcharting Standards

Just as definitions of words are necessary for communication, agreement on the meaning of flowcharting symbols is important. To date, considerable progress has been made in the standardization of flowcharting symbols as a result of the work of the American National Standards Institute (ANSI). The ANSI standards have been widely accepted and are employed by most professional systems analysts.

The student should become totally familiar with these ANSI standard flowcharting symbols. Using them helps to maintain the high standards of the systems profession.

Input/Output Symbol

This generalized input/output symbol may be used to represent either input data or output when the medium (such as magnetic tape, card, paper form) is not designated. It is particularly useful in the early stages of systems analysis before input/output media have been determined.

Punched Card Symbol

This symbol is used when either input or output data is in the form of a punched card.

Document Symbol

This symbol may be used to represent original source documents, such as employee time cards which serve as input to a system. It may also be used to represent printed output, such as computer-prepared reports which are produced by a system.

Magnetic Tape Symbol

This symbol is used to represent input or output data which is being stored on magnetic tape.

Process Symbol

This symbol is used to represent a computer program or some other machine operation or series of operations.

Manual Operation Symbol

This symbol represents a manual operation or some auxiliary machine operation which requires an individual operator, such as data entry.

Online Storage Symbol

This symbol represents data which is being stored on a direct-access device, such as a magnetic disk or drum.

Punched Tape Symbol

This symbol represents input or output which is being stored on punched paper tape.

Offline Storage Symbol

This symbol denotes that data is being stored external to the computer system. It is a generalized symbol in the sense that the stored data may actually be on card, tape, or disk packs.

Manual Input Symbol

This symbol shows that input data is being supplied directly to a computer from an online device, such as a computer terminal or console.

Communication Link Symbol

This symbol shows that data is being transmitted automatically from one location to another via a communications network.

In addition to the above symbols which may be used in both systems and program flowcharts, there is another called the decision symbol, which is more frequently used by programmers.

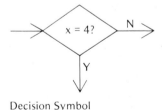

Decision Symbol

This symbol is generally posed as an unanswered question or variable condition, for example, x = 4. If the condition is met, one path (Y) will be followed. If it is not met (N), a different direction of flow is followed.

Reading a Flowchart

Once you understand the standard flowcharting symbols, reading a flowchart becomes a straightforward process. In figure 3-8, side notes explain what is being graphically illustrated by the flowchart.

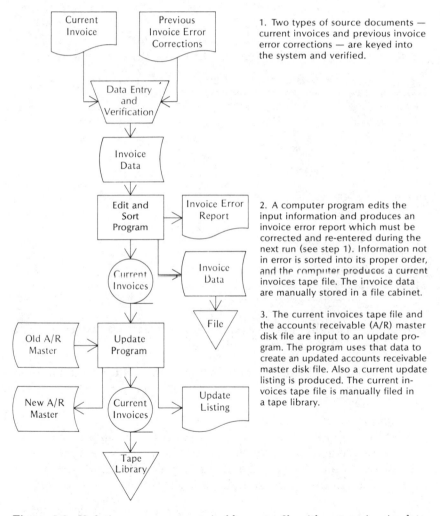

1. Two types of source documents — current invoices and previous invoice error corrections — are keyed into the system and verified.

2. A computer program edits the input information and produces an invoice error report which must be corrected and re-entered during the next run (see step 1). Information not in error is sorted into its proper order, and the computer produces a current invoices tape file. The invoice data are manually stored in a file cabinet.

3. The current invoices tape file and the accounts receivable (A/R) master disk file are input to an update program. The program uses that data to create an updated accounts receivable master disk file. Also a current update listing is produced. The current invoices tape file is manually filed in a tape library.

Figure 3-8   Updating an accounts receivable master file with current invoice data

DECISION TABLES

Decision tables are a means for representing decision alternatives in a tabular form. They are most effective when the problem is one of selecting a single decision alternative or set of alternatives out of many possibilities.

Results Tables

You are probably quite familiar with one elementary kind of decision table called a *results table*. The results table is the kind of two-dimensional table commonly used in train, bus, and airplane schedules, and for income tax computation. Table 3-1 is a simplified example of a results table that might be used for income tax com-

Table 3-1   Results table for income tax deduction

Net Taxable Income	Number of Dependents				
	1	2	3	4	5
Under $13,000	348	129	0	0	0
$13,000–$13,999	463	326	204	68	0
$14,000–$14,999	625	481	378	264	148
$15,000–$15,999	811	636	528	422	306
$16,000–$16,999	982	815	688	575	467
$17,000–$17,999	1257	978	859	740	621
$18,000–$18,999	1482	1149	1030	911	792
$19,000–$19,999	1729	1320	1201	1032	963

putation. Notice that in this example there are two sets of conditions (dimensions) affecting the base tax rate:

Condition 1.   The net taxable income range of the individual
Condition 2.   The number of dependents being claimed

To find the correct tax base you match the appropriate net income range to the appropriate number of dependents. Each possible set of two conditions and their result is called a *rule*. By using a results table we are able to express in a single table all of the rules for income tax calculation. An example of one of these rules would be:

Condition 1   IF the net taxable income range = $15,000–$15,999

Condition 2   IF the number of dependents = 4
    Result   THEN the tax base = $422

Usually the systems analyst must deal with decisions that involve many conditions. When more than two conditions are involved in a decision, the results table loses its effectiveness and a different kind of decision table must be employed.

The Quadrant Method

The quadrant method for constructing decision tables derives its name from the fact that it is structurally made up of four major parts, or quadrants, each of which has a unique name. Figure 3-9 illustrates the placement of each of the following four quadrants: (I) condition stub; (II) action stub; (III) condition entry grid; and (IV) action entry grid.

The condition stub contains a separate statement for each of the conditions which might possibly affect the decision outcome. These statements are generally formatted as IF statements.

The action stub contains a separate statement for each action which might possibly be required. The action statements are usually formatted as THEN statements.

Taken together, the condition and action statements present a traditional IF . . . THEN logical relationship. As we shall see in our later discussion of the uses of decision tables, this precise logical relationship makes decision tables a most effective tool in many instances.

Because a number of combinations of conditions and actions may exist, it is necessary to record all of the various possibilities directly on the decision table. The condition entry grid is used to record which conditions apply or do not apply. If a condition applies, a Y is entered onto the grid. If it does not apply, an N is entered.

Quadrant I	Quadrant III
Condition Stub	Condition Entry Grid
IF . . .	Y or N
Condition Statements	
Quadrant II	Quadrant IV
Action Stub	Action Entry Grid
THEN . . .	X or .
Action Statements	

Figure 3-9   Decision table quadrants

Title: Credit Determination Table								
Condition/Action Statement	\multicolumn{8}{c}{Decision Rules}
	R_1	R_2	R_3	R_4	R_5	R_6	R_7	R_8
If customer credit limit not exceeded	Y	Y	Y	N	N	N	Y	N
If customer pay history rated good or better	Y	Y	N	Y	N	Y	N	N
If customer credit rating good or better	Y	N	N	Y	Y	N	Y	N
Then approve order without stipulation	X	X	•	•	•	•	•	•
Then approve order pending receipt of cash	•	•	•	•	X	•	X	•
Then refuse credit	•	•	X	•	•	X	•	X
Then refer customer to credit manager	•	•	X	X	•	X	•	•

Figure 3-10 Completed decision table

The action entry grid performs a similar function in terms of whether or not a particular action is to be taken. However, in this instance an X is entered onto the grid for a required action while a dot (.) is entered when that same action is not required.

Let us review some of the highlights of figure 3-10 which is an example of a completed decision table.

Title The title of the decision table should clearly explain the purpose of the table.

Condition statements These IF statements must be clear and concise.

Action statements These THEN statements, again, must be clear in their meaning and concise in their content.

Rules Each column of condition grid and action grid entries is called a rule. A rule is simply a statement that under a given set of conditions, specific actions must follow.

Number of rules The number of rules which exist for a problem can be determined arithmetically as follows:

1. Count the number of condition statements (in this case, 3).

2. Use the number arrived at in step 1 as the exponent of the number 2.

3. Compute the number of rules ($2^3 = 8$).

In so doing, we have determined that there are eight possible unique combinations of conditions for which actions must be specified.

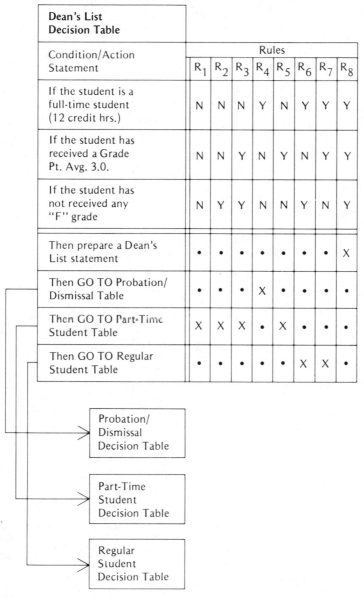

Dean's List Decision Table

Condition/Action Statement	Rules							
	R_1	R_2	R_3	R_4	R_5	R_6	R_7	R_8
If the student is a full-time student (12 credit hrs.)	N	N	N	Y	N	Y	Y	Y
If the student has received a Grade Pt. Avg. 3.0.	N	N	Y	N	Y	N	Y	Y
If the student has not received any "F" grade	N	Y	Y	N	N	Y	N	Y
Then prepare a Dean's List statement	•	•	•	•	•	•	•	X
Then GO TO Probation/ Dismissal Table	•	•	•	X	•	•	•	•
Then GO TO Part-Time Student Table	X	X	X	•	X	•	•	•
Then GO TO Regular Student Table	•	•	•	•	•	X	X	•

Probation/ Dismissal Decision Table

Part-Time Student Decision Table

Regular Student Decision Table

Figure 3-11 Multiple decision tables using a GO TO statement

Multiple Decision Tables

At times complex business decisions are more properly viewed as a set of related decisions as opposed to a single decision. Each one of these subset decisions may be sufficiently complex to warrant its own decision table. At such times it is possible to develop several decision tables and to include statements which permit interaction between the various tables. For example, the multiple decision table in figure 3-11 illustrates the use of a GO TO action statement. This GO TO statement directs the user to the next decision table for consideration.

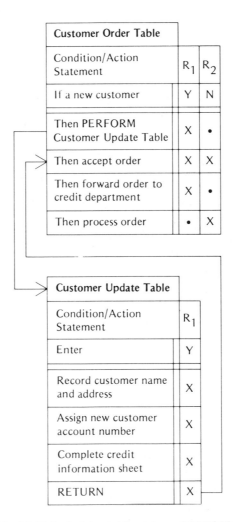

Figure 3-12   Multiple decision tables using a PERFORM statement

A second example is given in figure 3-12, where the PERFORM action statement is also employed. The PERFORM action statement directs the user to some other table but with the stipulation that a RETURN be made to the first table once all of the actions contained in the second table have been completed.

Note that decision tables which are performed are somewhat different from others. Instead of condition statements they simply have an ENTRY point. Also, the last action to be performed must be a RETURN action statement. This RETURN statement points back to the original table. Upon returning to that table, one begins with the action statement which immediately follows the initiating PERFORM statement.

Decision Table Applications

As a systems tool, decision tables work best in those instances which are both specific and detailed. For example, using a decision table which clearly outlines the credit action to be taken under specific conditions, a clerical worker with little or no training can accurately process or refer numerous credit applications.

Similarly, the systems analyst can provide decision tables to computer programmers as a basis from which coded routines can be developed. Such an approach eliminates much of the guess work from programming and can particularly assist the less accomplished programmer to do better programming work.

Decision tables are also a beneficial addition to systems documentation. By including decision tables in the formal documentation, the analyst can quickly review some of the complex decision making processes without wading through computer program listings or lengthy, written statements and procedures.

THE QUESTIONNAIRE

The questionnaire, or *survey questionnaire* as it is sometimes called, is a tool which enables the systems analyst to record people's opinions in a uniform manner. There are essentially two types of questionnaires:

- The open-ended (or free-format) questionnaire
- The fixed-format questionnaire

While both types of questionnaires are useful, the fixed-format questionnaire has the advantage of lending itself to rapid machine

tabulation. For this reason, it is the type of questionnaire most frequently employed, particularly when large numbers of people are
being surveyed.

Constructing, issuing, and evaluating questionnaires is a topic
for intensive and extended study. Here, we will make only some introductory comments on constructing, issuing, and evaluating questionnaires.

Free-Format Questionnaires

The free-format questionnaire is one which basically poses an unanswered question.

Example 1
What kind of computer do you currently have installed?

Name	Model	Quantity
_____	_____	_____

Example 2
Give the name and model number (for example, IBM 4341
group II) and quantity of all computers

Currently installed: _____

Currently on order: _____

The format used in each of these examples is essentially that of an
unanswered question, the answer to which can be made: (1) quickly,
(2) accurately, and (3) in brief, common terminology. If the person
responding to the questionnaire (respondent) cannot answer
quickly, we will probably not receive a response. If he or she cannot
respond accurately, we do not really want the response. And if the
response is not in brief, common terminology, we will have difficulty
quantifying and perhaps understanding the response. A well-
written free-format questionnaire meets the above criteria.

Fixed-Format Questionnaires

The fixed-format questionnaire is a different approach to the search
for information. It presents either a statement or a question regarding the problem, and a limited number of fixed responses from
which the respondent must choose.

Example 1
This company's experience with the use of computers as a means for assisting in inventory control has been successful.

☐	☐	☐	☐	☐
Strongly Agree	Agree	No Opinion	Disagree	Strongly Disagree

Example 2
Do you feel that your company's experience with the use of computers as a means for assisting in inventory control has been:

☐	☐	☐	☐	☐
Very Successful?	Successful?	No Opinion	Unsuccessful?	Very Unsuccessful?

There are some major considerations required in the construction of a fixed-format questionnaire. First, it is probably best to limit the scope of the questionnaire by limiting the number of specific topics being investigated. This enables the analyst to direct the questionnaire to a specific group. If many topics are covered in a single questionnaire, it may be difficult to find individuals who are able to provide accurate answers to all of the questions. This may mean that the analyst will have to contact several individuals to complete a single questionnaire.

It is also wise to avoid general questions. A number of specific questions about the general topic will help the analyst to obtain a more flexible, indeed a more meaningful, response. For example, if, instead of asking a single question about the successful use of computers for inventory control, several questions had been developed regarding specific areas, it might have looked like this:

Example 3
Do you feel that the use of computer assistance for inventory control:

(1) has improved the ability of your company to respond to customer orders and/or customer requests?

☐	☐	☐	☐	☐
Strongly Agree	Agree	No Opinion	Disagree	Strongly Disagree

(2) has been effective in helping to manage your company's finished goods inventory?

☐ ☐ ☐ ☐ ☐

Strongly   Agree   No   Disagree   Strongly
Agree   Opinion   Disagree

(3) has had a positive effect upon production planning activities?

☐ ☐ ☐ ☐ ☐

Strongly   Agree   No   Disagree   Strongly
Agree   Opinion   Disagree

(4) has proved to be of valuable assistance to sales personnel?

☐ ☐ ☐ ☐ ☐

Strongly   Agree   No   Disagree   Strongly
Agree   Opinion   Disagree

Needless to say, the four questions given above do not even begin to cover the topic of computerized inventory control. However, by including clusters of questions for a small number of topics: (1) the questionnaire becomes more flexible since people are able to respond to the particulars of what they like and dislike, and (2) the systems analyst is able to make more refined judgements about people's opinions.

Structural Considerations

In the previous examples of fixed-format questionnaires, an attempt was made to balance the affirmative against the negative responses which could be selected and to further allow for self-disqualification through the "no opinion" option. The reason for allowing self-disqualification should be quite obvious. That is, uninformed, undecided, and disinterested responses are not being sought, and this option provides a polite way out for the person who is responding to the questionnaire. The reason for balancing alternatives is perhaps not so obvious.

Had the responses been limited to four alternatives, namely: (1) strongly agree, (2) agree, (3) no opinion, (4) strongly disagree, the responses themselves may have prejudiced the final results. This could have occurred quite naturally since the four alternatives allow for two positive responses (1 and 2) and only one negative response (4). To help avoid built-in bias, the proper balancing of fixed alternatives is a good practice.

Lastly, a possibility which should be considered in the construction of any questionnaire is that the respondent has not, for one reason or another, responded fairly and honestly. It may be that in haste or disinterest the respondent has only politely, and not accurately responded. It may also be true that the questionnaire was not truly relevant, or did not allow for the freedom of expression which the respondent desired. To cover these and other possibilities, it is advisable to include questions which allow for polite self-disqualification from the entire questionnaire. This is normally accomplished by either asking for a specific evaluation of the relevance of the questionnaire or by addressing oneself to a hypothetical situation with which the respondent can possibly identify. For example:

Example 1
Although interesting, the questions asked above do not directly relate to this company's use of computers for inventory control.

☐   ☐   ☐   ☐   ☐

Strongly   Agree   No   Disagree   Strongly
 Agree   Opinion   Disagree

Example 2
Although interesting, this questionnaire comes at a time when business priorities prevent me from giving complete attention to it.

☐   ☐   ☐   ☐   ☐

Strongly   Agree   No   Disagree   Strongly
 Agree   Opinion   Disagree

It is a matter or choice as to where these kinds of questions appear in the questionnaire. Some prefer to intersperse them while others choose to place them at the end of the questionnaire. In any event, when these kinds of questions are responded to in the affirmative, you can put little trust in the accuracy and reliability of the questionnaire results.

Issuing Questionnaires

The first important step in issuing a questionnaire is to understand what groups (populations) are to be surveyed and to randomly select individuals who are representative of their groups. If, for example, one is attempting to sample opinion on a college campus, it is important to know if the individual is a member of the board of trustees, administration, faculty, or student body. Also, it is necessary to survey the group on a proportionate basis. Since there are generally more students than faculty on a campus, we would expect the total sample to include proportionately more students than faculty. If many different variables must be accounted for, then representative, random sampling can become a very involved and time-consuming process, the full discussion of which is not within the scope of this chapter. However, it should be noted that the interpretation of data obtained through the use of questionnaires can be adversely affected by not adhering to proper sampling procedures.

One obvious advantage of a questionnaire is that it can be issued in a number of different ways by people who have received only limited training and who know little about the problems being discussed in the questionnaire.

The simplest and least expensive way to issue a questionnaire is simply to mail it out to those people whose opinion is being sought. As usual, this easiest approach is not the best. Questionnaires that are mailed out frequently find their way into the wastebasket; thus the rate of return is characteristically low. Equally important, those who respond to mailed questionnaires may not be representative of the total group being surveyed. Frequently, it is the person with an extreme point of view (either positive or negative) who will feel compelled to return a questionnaire which he receives through the mail. While these extreme viewpoints are important, they should not be the sole basis for a decision.

When the person to be surveyed is physically distant, a reasonable alternative to mail-out questionnaires is to issue the questionnaire via the telephone. In this way it is usually possible to find a time which is mutually convenient to complete the questionnaire. However, problems may still exist in attempting to issue questionnaires in this manner. If you are the person placing the telephone call, respondents may want to know more about you and the organization you represent before they are willing to talk freely on the telephone. This may require a letter to respondents from a high official within your organization explaining the importance of your work and requesting the respondents' cooperation.

Whenever possible, direct personal issuance of the questionnaire should be attempted. Issued on a one-to-one basis, the questionnaire is readily completed, and a high degree of confidence can be placed in the accuracy of the responses. However, the following guidelines should be followed when issuing a questionnaire in this manner:

1. Read a standard well-written introduction which clearly states:
   a) Who you are (name, position, and organization)
   b) The purpose and nature of the questionnaire
   c) On what basis the respondent has been selected (for example, as a director of marketing)
   d) The format in which the response should be made
   e) An estimate of the amount of time that will be required to complete the questionnaire
2. Read each question to the respondent and if a fixed-format questionnaire is used, also read the alternatives from which the respondent must choose.
   a) Read all questions clearly and evenly.
   b) Avoid voice inflections and gestures which might encourage one response rather than another.
3. React to questions by rereading the statement a second or third time if necessary.
   a) Avoid ad lib interpretation of the statement for the respondent.
4. Record all responses accurately. In addition to checking the proper choice on a fixed-format questionnaire, pertinent comments made by the respondent should be entered directly on the questionnaire form.
   a) These notes are particularly important as they relate to the respondents' interpretation of the question. If the question is interpreted differently by different people, then these notes will become the basis for eliminating the question from the final tally.

Quantifying and Interpreting Results

If proper attention has been given to the construction of a questionnaire and its issuance, this phase of the work is quite straightforward. If a fixed-format questionnaire was used, the process is further simplified because the analyst can use the computer in the tabulation and interpretation of results. In any event, it is worth noting the following in any formal written presentation of the results:

1. The purpose and nature of the questionnaire (including a sample copy of the questionnaire form)

   2. The populations or groups or people surveyed
      a) Some comment on the numbers surveyed for each sub-group
         within the total population
      b) Some comment on the basis which was used for establishing
         a representative sample

   3. Results by population sub-group and in total, including:
      a) Number of self-disqualifications from the survey
      b) Number of responses to each question by response category
         (strongly agree, agree, and so on). These figures should also
         be stated as a percentage of all responses to that question
         (for example, 58% agree).

   4. Conclusions should be based upon results which:
      a) Pinpoint areas of potential conflict, shown by strong differ-
         ences of opinion between people
      b) Clearly indicate actions which should be taken
      c) Raise questions regarding the validity of portions of the
         questionnaire, stating the reasons
      d) Imply areas of uncertainty requiring further investigation

The analyst who decides to use the questionnaire should also plan
sufficient time to make a formal presentation such as that outlined
above. In this way there is an improved possibility that the survey
findings will be properly interpreted. In some instances it will also
help to extend the distribution of the survey findings to others who
may find them valuable. When a formal presentation of results is
prepared, it is a generally accepted policy to distribute a copy of the
presentation, or a summary of the results, to all parties who partici-
pated in the survey.

CHAPTER SUMMARY

Six basic systems tools were discussed in this chapter: the HIPO
technique, data flow diagrams, N-S charts, flowcharts, decision
tables, and questionnaires.
   The HIPO technique is a structured tool with broad applicability
throughout the systems development life cycle. This technique uses
two different types of charts. The first type, called a VTOC diagram,
is used to develop a top-down hierarchical structure through which
systems functions and their relationships can be observed, discussed, and better understood. The second type is frequently called
an IPO diagram, because it describes the input, output, and process
components for a given function.

The data flow diagram is also a structured technique and is frequently used as an alternative to an IPO diagram. Data flow diagrams are easy to draw and require the knowledge of only a few symbols. However, care must be taken to avoid making them overly detailed.

N-S charts have been primarily used to design structured programs. However, they should prove helpful to the analyst in clearly describing complex problems in systems logic.

Systems flowcharts provide a concise diagram of a complete system or process. Thus, they can be useful in problem organization, problem solving, problem presentation, and problem review.

Decision tables represent human or machine decision alternatives in tabular form. They are most effective when the problem is a complex decision with many different conditions and possible results.

The questionnaire or survey questionnaire is used by the analyst to sample other people's opinions in a uniform manner. To be effective, care must be taken to construct, issue, and interpret questionnaires scientifically.

In the future, you can expect continuing and expanding use of the structured tools discussed in this chapter. You can also expect continuing use of more traditional tools, such as flowcharts and decision tables. However, the application of these traditional tools will increasingly be within the context of an overall structured, top-down approach.

WORD LIST

action entry grid—that quadrant of a decision table which specifies which actions are or are not required

action stub—that quadrant of a decision table which contains the action statements

condition entry grid—that quadrant of a decision table which specifies which conditions apply or do not apply

condition stub—that quadrant of a decision table which contains the condition statements

data flow diagram—a structured technique that is effective for logical systems design, because it uses symbols that do not impose physical constraints

decision table—a means for representing human or machine decision alternatives in a tabular form

fixed-format questionnaire—a questionnaire which poses a series of statements or questions and a limited number of fixed responses from which the respondent must choose

free-format questionnaire—a questionnaire which poses a series of unanswered questions to which a person can respond in his or her own words

HIPO—an acronym for hierarchy plus input-process-output, which is a structured technique with broad application throughout the systems development life cycle

IPO diagrams—charts used in HIPO that describe the inputs, processes, and outputs for a system

Nassi-Shneiderman (N-S) chart—a chart, named after its codevelopers, which is primarily used to design structured programs and is effective for working with complex logical problems

program flowcharts—standard flowcharts which present a detailed illustration of the steps involved in solving a problem

quadrant method—a method for constructing decision tables which will accommodate many sets of conditions and multiple results

results table—an elementary form of decision table which is limited to dealing with two sets of conditions and a single result

standard flowcharts—flowcharts which use standard American National Standards Institute (ANSI) flowchart symbols

structured, top-down analysis—an approach to systems analysis requiring that the structure of a system be developed first, beginning in overview fashion and working down to increasingly detailed levels

systems flowcharts—standard flowcharts which present a general overview of an entire system or process

visual table of contents (VTOC) diagram—a chart used in HIPO that reflects the hierarchical relationships among systems functions

QUESTIONS AND PROBLEMS

1. Construct a standard systems flowchart from the following functional description.

   Using a computer terminal at a remote location, an order clerk enters order data directly into a computer system, in which the following events occur:

   a. The order is priced by using the master price file (disk).
   b. The inventory is selected from the inventory master file (disk).
   c. The customer's name and shipping address are retrieved from the customer master file (disk).
   d. A customer invoice is printed, and a customer invoice tape is produced.
   e. A confirming notice that the order has been processed is sent to the order clerk.

2. Slowshift Transmission, Inc. offers a comprehensive employee benefits program, including health (Blue Cross/Blue Shield and major medical and dental coverage), retirement, tuition support, and college donation matching.

   Construct an IPO diagram for the donation matching program that illustrates:

   a. How the employee contributes to a given college and at the same time registers with the benefits office by completing form B21-4

   b. How the college notifies the employee and the firm that the donation has been received

   c. How the firm sends payment to the college and notifies the employee; be sure to illustrate how problems will be dealt with (for example, the firm may receive notice from a college for which no form B21-4 is on file)

3. For the following flowchart, provide a functional description that illustrates your knowledge of standard flowchart symbols:

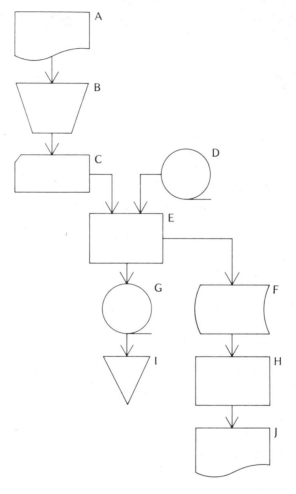

4. Big Apple Garden Supplies Company has been having difficulties with its warehousing operation. Normally, the retail outlet fills customer orders from its own inventory. However, occasionally a customer wants an item that is not in stock. At such times, the order is placed at the warehouse for direct shipment.

In the usual procedure, the retailer places bulk orders with the warehouse. The warehouse is accustomed to handling these bulk orders and seems to have little trouble with them. However, the warehouse has not been so successful in handling direct orders.

As the systems analyst assigned to the project, you are to:

a. Construct an overview process flow diagram that illustrates the interaction among customer, retail outlet, and warehouse

b. Describe the advantages of using this structured approach

5. Using the quadrant method, construct a decision table for the following conditions for granting dean's list status:

If a student has at least a 3.0 grade-point average for twelve or more credit hours during a given semester and did not receive any failing or incomplete grade during that semester, then the student is entitled to dean's list status for that semester.

6. You have been asked by your manager to prepare a presentation on the structured, top-down approach. Using the technique of your choice, prepare a presentation in which you:

a. Explain the use of this approach with illustrative diagrams (for any system you want)

b. Describe this approach's advantages over nonstructured approaches (and over other structured approaches, as appropriate)

MINICASE—CONTEXT INTERNATIONAL, LTD.

Your work as a systems analyst with Context International has caused you to conclude that:

1. The payroll function must be standardized for each of the subsidiary companies. Time reporting, paycheck preparation, job accounting, and subsidiary management reports need common coding and processing structures. The results of this process must be transmitted to corporate headquarters for federal reporting and comparative analysis.

2. The accounts payable function must be expanded to integrate purchasing, payables, accounting, and check writing functions.

3. The accounts receivable function must be expanded to integrate customer order credit checking against outstanding receivables, invoicing, cash application, trial balance, and dunning processes.

4. The cost accounting function must focus more clearly on job order costing. A system is required that will use materials and labor information, as well as overhead costs, to determine the actual cost of goods sold.

5. The general ledger function must be expanded to include a corporate-wide consolidation process. This expansion will require that subsidiary general ledger information be transmitted to the corporate headquarters, where the information will be consolidated to produce a corporate-wide income statement and balance sheet.

Based on the limited information provided above, prepare the following:

- A VTOC diagram for a new financial system
- A data flow diagram for the payroll subfunction
- An IPO diagram for the general ledger subfunction

NOTES

1. See the latest IBM HIPO reference manual for a more in-depth description of the HIPO technique.
2. C. Gane and T. Sarson, *Structured Systems Analysis: Tools and Techniques* (New York: Improved Systems Technologies, 1977). This general reference book emphasizes the use of data flow diagrams.

4

Problem Definition
and Classification

CHAPTER OBJECTIVES

When you finish this chapter you will be able to—

- *Discuss the issue of gaining organizational commitment.*
- *State and discuss those major systems problems most frequently encountered.*
- *Identify major systems problems from a narrative description of a problem situation.*
- *List and discuss the levels of systems problem complexity.*
- *Identify and discuss the kinds of conflict which are frequently encountered in problem definition and classification.*

The materials covered in this chapter and those in the following chapter on data collection and analysis are highly related and interdependent. The reason is that, in order to properly define and classify a problem, it is necessary to collect and analyze considerable amounts of data.

While it is true that considerable data must be collected and analyzed before a problem can be completely defined, the analyst must have some direction to the search for relevant information. So, the analyst begins almost immediately to define and classify the problem. These early tentative conclusions direct the search for additional information and are modified as new evidence is uncovered. Through this process of redefinition, the analyst forms a clear statement of the problem and a detailed understanding of why the problem exists.

A clear, in-depth understanding of the problem provides the framework within which the systems process will unfold. Arriving at this point will be both difficult and time-consuming. Nevertheless, by taking the time early in the systems process to carefully and thoroughly investigate all aspects of the problem, the analyst improves the potential for success in finding the "best" solution to the problem.

Problem definition and classification refers to something which is considerably more complex than simply knowing that a problem exists. In fact, to know that a problem exists is not the primary function of systems analysis. Rather, it is the starting point. The following simplified example illustrates how a problem might be detected and how the mechanism for systems analysis action might be set into motion.

Company XYZ is a manufacturer of tennis balls. In addition to its use of independent auditors, management has in recent years developed a small team of internal auditors whose job has been the reviewing of the financial operation and management of the company. During a recent audit of the Payroll Department, that department was cited by the auditors on the following major points:

1. There were some discrepancies between timesheet information and paycheck amounts.
2. Insufficient information was being maintained to permit a thorough audit of past transactions.
3. Insufficient written documentation was available regarding payroll procedures and personnel duties and responsibilities.
4. Insufficient means were available for providing internal security posing a real possibility that fraud or embezzlement could occur undetected.

As a part of its recommendation, the internal audit team recommended that a number of immediate corrective steps be taken by the Payroll Department. It also recommended that management consider an in-depth analysis of the Payroll Department with an eye towards reorganization of the department and redesign of the payroll system.

In the above example systems analysis was viewed as an important approach used in dealing with known business problems. Having said this, it is important to quickly add that although it was known that a business problem existed, it was not known what kinds of systems problems caused or contributed to the difficulties. Thus, the initial efforts of the analyst would be directed to the task of performing an in-depth definition and classification of the systems problems.

In problem definition and classification, the analyst must:

• Obtain organizational commitment
• Determine the kinds of systems problems that exist
• Ascertain problem complexity

The definition and classification of the problem must be performed in a way which lends itself to further systems development, design, and implementation. In other words, the analyst must consider early findings in the light of what can reasonably be done to correct the problem. This point will hopefully become more apparent in the following discussions.

GAINING ORGANIZATIONAL COMMITMENT

For the analyst to obtain the organizational commitment necessary to effectively deal with a systems problem, the organization must view both the problem and its solution as important. Clarifying the degree of organizational commitment permits the analyst to better understand the resources and support that can be expected.

It should be noted that in asking how important a problem is, you are in effect asking to have a priority placed on solving the problem. If we view the resources which are available to an institution as essentially a "scarce" commodity then we will also expect that there will be some internal competition and even conflict over who will have access to those resources. In other words, the question of problem importance is frequently characterized by much debate. Ultimately, the answer which is obtained will reflect the level of resource commitment which the analyst can expect to use in solving the problem.

In attempting to gain the required degree of clarification and commitment, the analyst needs to:

- Obtain management's assessment and involvement
- Assess user influence and conflict
- Gain the active participation of the right people
- Develop a strategy for maintaining management and user involvement and commitment

Each of these important issues will be more fully discussed in the following sections.

Management Assessment and Involvement

As was briefly mentioned in chapter 1, the role of management is crucial to systems analysis. It is through management that the analyst gains an understanding of institutional goals and objectives.

Management's view of the relative importance of a problem will to a large extent depend upon the direct relation of the problem to institutional goals and objectives. Consequently, the analyst must present the problem to management in a manner which makes this relationship clear. For example, to outline a situation emphasizing the frustrations of workers who are attempting to make important decisions without proper information is insufficient to gain management consensus that an important problem exists. From management's point of view it may simply appear that they have some "bad" employees.

If, on the other hand, the lack of available information can be tied to poor customer service or higher inventories, then the problem becomes critical from the managerial perspective.

In the same way that some problems become priority concerns because they are relevant to management goals and objectives, others are deemed less important either because they are not or because management does not choose to view them as relevant. This point is mentioned because it does happen that even the most relevant matter can be ignored simply because management does not choose to give it legitimacy. It is the professional obligation of the analyst to bring important problems to the attention of management. However, it is management's prerogative to encourage or constrain the systems effort based upon its assessment of relative importance.

Management—and particularly top management—is itself a scarce resource, in that the time the top manager can afford to give to any one person or issue is very limited. Thus, as a first premise, the analyst must remember not to waste the time that is made available. The time should be used to transmit important, management-related information concisely and clearly. This approach in itself helps achieve management involvement, because the manager will, over time, learn that involvement will not require spending vast amounts of time on petty issues.

A second important aspect of gaining management involvement is to give the manager feedback that illustrates how previous involvement proved important to achieving success. Too often, top managers hear from others only when there is a crisis or an urgent need. In these instances, they may commit resources to deal with the problem, but they may learn about the outcome only second-hand. Management does not appreciate receiving this kind of feedback and may, as a result, choose to be less involved or helpful the next time. Do not make this mistake. If management commitment and involvement are obtained, then you must give management appropriate feedback about the outcomes. In so doing, you also create an opportunity to discuss other management involvements that might be most appropriate and helpful in the future. Such an approach creates a basis for maintaining an ongoing dialogue with management.

Finally, the wise analyst is fully prepared for encounters with management. The analyst attempts to anticipate the kinds of questions that might be asked and has materials ready that are pertinent to those questions. Managers prefer to place limited resources where they will achieve the greatest gains. Their perception of the analyst as a thorough, competent, and prepared person is established in large part through direct encounters—however few those may be.

User Influence and Conflict

The question of "How important a problem?" frequently is not resolved by management alone. There are many people and groups of people within the organization who may hold differing opinions about the problem and who have at least some influence on the priority which is given to resolving the problem. Let us briefly sketch two different hypothetical situations.

Situation 1:
There are four major functional areas in Company Y. Three of the four areas feel that they could benefit from an extended area systems analysis. The company's systems department is small enough to preclude any possibility of studying more than one area at a time. The company's financial picture is bleak, so there is no possibility for increasing the size of the systems staff and no way for it to meet the needs of all three areas simultaneously.

Given the above situation, we can anticipate conflict. Each area seeking systems help will use its influence to gain support for itself (perhaps even if it only means that in the short run no other area gets that support). It is likely that they will put pressure on uncommitted parties such as the systems department, the fourth area, and management in the struggle to win.

Situation 2:
A problem has been diagnosed and defined in Office ABC. There are a number of other offices that depend upon Office ABC for information. The other offices stress the importance of the problem relative to their operations. The director of Office ABC feels that the importance of the problem has been greatly overemphasized. The director also feels that this problem has been used as a cover for other far more important problems in the other offices. The director also feels that specific difficulties involved in dealing effectively with the problem will create more problems in his department rather than fewer.

Knowing that many people see the existence of a problem as a kind of personal slur, and also knowing that there is a great deal of scapegoating which routinely occurs in organizations, the analyst must remain as objective as possible in situations like this. It appears to be in the very nature of organizations that they are essentially political at times and, therefore, subject to conflict.

The above situations represent only two of a virtually unlimited variety of possible conflicts. What is important to note here is that the question of "How important a problem?" frequently results in the creation of internal conflict and the use of influence as a means for resolving it.

Because conflict creates, at times, a less than pleasant environment, some people are prone to avoid it. A conflict-fearing analyst might attempt to propose only those things which people do not feel strongly about, or to do things "behind the scenes," bringing them out in the open after-the-fact. It is a major premise of this book that systems analysis must be an open and honest process which has as its primary focus major systems problems. Conflict is viewed as natural and can be beneficial. It allows many to participate, invites resolution by the process of compromise, delineates the practical limits of what can be done (which we call a "settlement" or "contract"), and provides the support basis for doing the agreed-upon job. Conflict will help to ensure that the important problem is labeled *important,* even after the test of debate. Ultimately, there will be a clearer understanding of why a particular project is being undertaken, and the resource commitment needed to do the job effectively will be there.

There are a number of rational steps which the analyst can take in attempting to:

- Eliminate conflict which results from nonlegitimate sources, such as scapegoating, covering up, and bias.
- Develop a scheme which channels conflict in a positive direction.
- Resolve conflict in a manner which results in a consensus of what should be done.

In a larger sense it is the structured involvement of users which affords the analyst the greatest opportunity for handling conflict in an effective manner. Whether it is a committee, a task force, or specific individuals, good decisions require the interest and imagination of many people. By working with conflict, the analyst will hear and examine many diverse opinions. The resolution will be, in large part, the result of the mutual education that occurs between parties during conflict

Getting the Right People Involved

Determining who are the right people to involve in a systems project can be a complicated task.[1] Naturally, we want the involvement of management and of people who are affected by the systems problem. However, there are also other guidelines that the analyst can use in evaluating which persons in those groups would make the greatest contributions as team members if they were involved.

There are three important sets of criteria for evaluating prospective team members. The first of these, called the individual perspective, stresses the selection of team members on the basis of individual characteristics, including general intelligence, dependability, social sensitivity, and good adjustment.[2]

The second set pertains to how well the individual works with others. This perspective stresses selecting people with strong interpersonal skills who also value the project goals and share in the team members' specific work objectives.[3]

The third set relates to organizational considerations. In this perspective, the ideal participant is one who has sufficient authority to make the necessary decisions and commitments, influence within the total organization, and project-related knowledge and skills.[4]

It is not always possible to identify individuals who are ideal according to all the above criteria. However, since all these criteria are important to the ultimate success of the project, it is best to try to achieve a balance within the team by including at least some members who are strong in each of these areas.

Strategies for Gaining a Continuing Commitment

In chapter 1 we presented the Steering Committee Model as one frequently employed method for determining how the Systems Department should interact within an organization. While the committee approach is not always "best" in the sense of guiding a particular systems project, it can be very effective in gaining a realistic level of commitment for the overall systems effort. There are several important reasons for this:

1. The committee approach helps to keep people informed as to the exact nature of the problem as well as the purpose and scope of the project.

2. The committee approach provides the opportunity to openly state either objections to the project or rationale to support it.

3. Committee decisions usually carry with them a statement of the expected level of organizational commitment associated with a given project.

By gaining some level of organizational commitment regarding the relative importance assigned to a problem area, the analyst is freed from the burden of having to justify to some people why "so much time" is being spent on a particular project, and to others why "so little time" is being spent on it.

It is possible to combine the Steering Committee Model with the Team Model (which was also mentioned in chapter 1), because the Team Model creates a work group whose commitment is more full-time and specific to the task of dealing with a particular problem, and not with all problems. Thus, the two models can complement each other. Because both are formal organizational structures, they can help ensure that management and user commitment will be sustained throughout the project.

An area of interest and concern pertaining to the extent of management and user involvement is part-time versus full-time participation. There are several alternatives that might be considered:

1. Full-time assignment of user personnel to the systems project team (perhaps even as project leader)
2. Full-time assignment of one or more people to act as liaisons between the project team and the user department
3. Part-time assignment of user personnel to function in a liaison capacity
4. Development of a committee or task force which will work with the systems team throughout the project

It has been observed that a small group of full time team members supplemented occasionally by part-time members with special expertise may make the most effective team.[5] Therefore, it is suggested that this approach be followed whenever possible.

WHAT KIND OF PROBLEM?

It is not uncommon to hear statements like, "We have an equipment problem," or "We have a personnel problem." These kinds of over-simplifications are not systems definitions. In many instances they are symptoms of a problem and not the real causes. For example, there are many potential reasons for having problems with people, including:

1. Do people have clear guidelines as to what must be done?

2. Are people properly trained to perform the job?

3. Do people have proper guidelines and supervision for performing the work accurately?

4. Do people understand the conditions under which problems should be referred to a higher authority?

5. Is there an efficient procedure available for dealing with job-related problems?

6. Is the process for evaluating the work performed by employees fair, accurate and reliable?

In addition to those items listed above, there are many other job-related factors which might have affected personnel performance. The point to be made here is that the systems analyst must take care not to prejudge what kind of problem exists, nor to oversimplify the definition of the problem.

The classification of the "kind of problem" should be done in a way which assists the systems analyst in the investigation by:

1. Directing the search for information
2. Encouraging objective assessment of the problem
3. Providing some indication of the kinds of technical and non-technical complications which can be anticipated

There are a number of kinds of problems that occur frequently within organizations. The following discussion of these problem categories is by no means exhaustive, but it does include some attention to most of them. The major kinds of problems to be discussed include:

• Problems of Responsiveness
• Problems of Throughput
• Problems of Economy
• Problems of Accuracy
• Problems of Reliability
• Problems of Efficiency
• Problems of Security
• Problems of Information

In any given situation one or more of the above problems can be involved. Also, proposed solutions for one kind of problem can, at times, have important consequences elsewhere. While recognizing that these interactions can and do occur, you will discover unique problem features if each kind of problem is discussed separately.

Problems of Responsiveness

The term *responsiveness,* as it is used in this section, means the speed with which a response can be made to a request for information or other service. For example, when a customer inquires about the status of a particular order, the responsiveness of the order department is based upon its ability to provide a quick response to the request for information. Naturally, there are times when the existing speed of response presents a major problem, and therefore must become the directive force for the systems investigation.

In dealing with this kind of problem the analyst generally orients the investigation around three major questions:

1. What is the current level of responsiveness and what contributes to it?

2. What can be done to streamline the existing organization and use of human and machine resources to attain greater speed and without sacrificing other standards of accuracy, economy, and so forth?

3. What additions or modifications in human and machine resources can be proposed to deal with the problem; and what will be the projected impact of these changes on the organization?

Clearly these different lines of investigation are necessary, and it is likely that they can be investigated simultaneously.

To a large extent, the analyst will be guided in this work by management's assessment of what constitutes the most satisfactory level of responsiveness. For example, management may believe that the optimum speed of response is to have an immediate answer to the customer's inquiry about the status of an order.

To avoid future misunderstandings, it is important that the analyst clearly state the requirements for improved responsiveness in quantifiable terms in any proposal for dealing with the problem. In this instance it would be appropriate to present a number of alternative solutions in which several different "responsiveness improvement plans" were fully explained in terms of their potential impact upon the organization.

Problems of Throughput

The term *throughput* has gained wide use among data processing and systems personnel. It refers to the amount of start-to-finish workload typically performed by an operation. For example, we might say that the order department is able to process 500 orders per day as an indication of its throughput potential. Unlike a problem which is strictly one of responsiveness, the emphasis here is essentially on quantity. Stated simply, the problem is how to increase

An order clerk working with a computer terminal gives immediate response to customer inquiries

the amount of work which can be completed in a given period of time.

Given this kind of problem orientation the analyst investigates the problem area seeking answers to the following questions:

1. What is the current throughput potential and what are the factors contributing to it?
2. What, if anything, can be done to improve the level of productivity and the utilization of resources which may have a direct bearing on throughput levels?
3. What resource modifications and additions could be used to increase the throughput potential? What would be the impact of their use on the organization?

When involved in this kind of problem, the analyst should be prepared to deal with the matters of workload standards and workload distribution. When we talk about standards we are really asking what is a "reasonable" amount of workload to be assumed by a particular person or piece of equipment. Naturally, this will be deter-

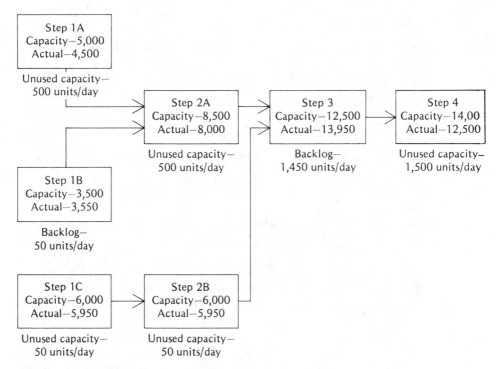

Figure 4-1 Workload distribution chart showing workload capacity and actual
workload received at each point

mined differently by different people in different organizations. The
analyst must help to define what is reasonable in the specific problem situation and then to equate what is being done to that standard.

Workload distribution is a second important consideration in
studies of this type. Like the weak link in a chain, the point of least
throughput often determines what the total throughput will be.
Therefore, the systems analyst must evaluate the throughput
capacity and the actual throughput at various points in the process
under investigation. It is important to note where there are backlogs of work to be processed and where there is available, unused
capacity. Figure 4-1 is a hypothetical example of a workload distribution problem. In this example, it is apparent that at steps 1b and
3 the throughput is insufficient to process the actual workload. At
all other steps in the process there is at least some degree of unused
capacity. The task of the systems analyst here is to determine what
kinds of corrective actions can be taken to deal effectively with the
throughput problem by eliminating both backlogs and unused
capacity from the process.

Problems of Economy

When the problem is one of economy, we mean that the major concern is reducing the average cost of services. Because the cost of doing business is crucial to the whole question of a company's financial stability, the analyst must understand problems of economy. To this extent it is important for the analyst to understand which of the following is the intent of management:

- To increase productivity at about the same cost, or with only measured, well-justified increases in cost.
- To maintain existing production rates but at a lower unit cost rate.
- To select certain unprofitable operations for discontinuance or consolidation, thus cutting both cost and production.

Certainly, the kinds of alternatives which are explored by the analyst will vary according to which of the above is the intention of management. However, in any event the systems analyst's task in this kind of problem situation will be a difficult one, and at times unpleasant. One way of analyzing the problem is on the basis of cost per unit of workload. For example, at each step in the operation it is possible to summarize both the throughput and all costs which can be associated with performing the work. The following simplified formula illustrates this relationship:

$$\frac{\text{Cost of performing the work}}{\text{Number of work units performed}} = \text{Cost per unit of work}$$

$$\frac{\$10,000/\text{day}}{20,000 \text{ orders/day}} = \$.50 \text{ to process one order}$$

Areas where there are unusually high or low unit cost rates may be earmarked for very close analysis. In any event, through this kind of analysis basic information is made available which is required for dealing with problems of economy.

Problems of Accuracy

While accuracy is always a concern in developing or modifying a system, there are times when it is the primary purpose for the analyst's involvement in a project. For example, in the previous illustration of the XYZ Payroll Department audit, the auditors found discrepancies between timesheet records and paychecks. The analyst who is assigned to that project will have to find out how and why the inaccuracies developed, as well as to propose methods for improving the accuracy of the system.

In projects where the primary problem is one of accuracy, the analyst becomes very much involved with the whole process of step-by-step verification of results. It is not enough to simply analyze the final result and evaluate its degree of accuracy. Rather, a careful series of checks and cross-checks must be instituted at critical points in the process. Also, procedures for correcting errors as they occur are required. In approaching this kind of problem the analyst should:

1. Determine logical points in the system where error checks and cross-checks should be instituted.
2. Determine where and how corrective actions can best be taken.
3. Determine what impact undetected errors will have upon other points in the process and upon the final systems product.
4. Establish what constitutes acceptable levels of accuracy at each check point.

Problems of Reliability

There is frequently some confusion regarding the difference between the terms accuracy and reliability. Quite simply, accurate results are valid while reliable results are consistent or repeatable. When the problem is one of reliability, the orientation of the analyst is towards improving the consistency with which accurate results can be obtained. An example of a systems study in which reliability is a major concern would be one in which an extensive back-up system is required. For example:

Company ABC has installed a large central computer system. As customer orders are received at various locations around the country, they are sent directly to this computer through computer terminals and an advanced data communications system.

During the past two months of operation there have been four instances in which computer system failure has resulted in lost customer orders, as well as long delays in the processing of many other orders. Management has determined that some action must be taken now to improve the reliability of the system.

The task of the systems analyst in the above example is twofold:

- To improve systems reliability by preventing the loss of orders when the computer fails.
- To improve systems reliability by avoiding severe delays due to computer system failure.

In dealing with this kind of problem, the analyst investigates alternate ways for processing the same workload. Also, methods for detecting that information has been lost and for resubmitting "lost" orders must be developed.

Problems of Efficiency

To define the term *efficiency* in a way that is both exact and acceptable to all people is difficult if not impossible. Yet almost everyone in the business world agrees that efficiency is important. Managers insist that they are primarily interested in more efficient ways to do business, and systems personnel frequently argue that their activities result in meeting this management concern.

Why is efficiency both important and elusive? It implies a balance of concern for each of the previously described problem areas. Ideally, an efficient system would quickly process large quantities of workload at a low cost and do so both accurately and reliably. In practice, the analyst must understand what an acceptable balance is between these different factors, and what the guidelines are for arriving at it. In one instance, it may be necessary to consider economy, given a system with three-second response time to a customer's request for information. This is quite different from an economic consideration when the response takes two hours or two days.

For problems of efficiency, the analyst must be concerned with each of the above factors, both as they exist independently and also as they relate to and affect the definition of one another.

Problems of Security

Problems of security generally come up when attention focuses on:

- Protection of information
- Protection of individual rights

In a growing number of widely publicized cases, persons with access to computer files have committed fraud. This has led to increased concern for the protection of information. Frequently, it is the systems analyst who is assigned to investigate existing systems with an eye toward developing methods and procedures which offer needed protection of information. Instances of vandalism and sabotage of equipment and computer files have created a second aspect of information protection to be considered by the analyst. While each aspect to this problem has its own unique considerations, it is now generally common to review the total question of information protection and to devise ways which protect information from both violent and nonviolent abuse.

With the development of large data banks of information which are shared by various governmental and commercial organizations, a heightened public concern for the protection of individual rights has begun to emerge. Given this climate of public opinion and concern, the question of who should have access to what information is important and must be considered by the analyst. Indeed, the question is so important that we can expect an increasing number of legal decisions on the matter. These decisions plus high professional standards for ethical practices by systems analysts will form the framework for dealing with this kind of systems problem.

Problems of Information

Information is a precious resource. It is the basis for effective decisions at all levels within a business organization. Problems in which the primary consideration is the use of information usually relate to one or more of the following:

1. The organization of available information
2. Human access to necessary information
3. The inclusion of additional information which was not previously captured
4. The classification of information as to its relevance and overall importance

Ability to work effectively with this kind of crucial problem will require considerable knowledge of the various alternatives which exist for coding, storage, organization, and accessing of information. The discussions presented in chapter 9 dealing with management information systems, computer-based files, and data-base management systems will provide some materials which are essential to working with this kind of problem.

HOW COMPLEX A PROBLEM?

While it is difficult to obtain a complete understanding of problem complexity prior to an in-depth analysis of the problem, it is necessary to initially develop some estimate of problem complexity. By developing such an estimate the analyst is better prepared to:

1. Schedule people and other resources that will be required for the investigation.
2. Estimate the extent to which work in this problem area will have potential implications for other areas.
3. Determine what levels of management should be involved in or made aware of the project.
4. Determine the extent to which these various management levels should be closely involved in the project.

During the early stages of systems analysis there are many "unknowns" regarding problem details. Therefore, it is helpful to have a limited set of general guidelines which can be used to assist the analyst in judging problem complexity. The following four categories, or levels, of problem complexity are suggested:

- Business Activity Subsystem
- Business Function Subsystem
- Business Operation Subsystem
- Business System

The use of the term subsystem implies that business activities, functions, and operations exist as smaller systems within the total business system.

Before examining each of these levels of complexity in some detail, it should be noted that they are intended only as initial guidelines. After an in-depth analysis of the problem, the analyst can develop a more refined estimate of problem complexity.

Business Activity Subsystem

A business activity is a procedure or series of steps to be followed in performing a single task. Examples of this level of complexity include "taking a customer's order" and "making a customer credit decision." From the systems perspective, even this lowest level of problem complexity must be viewed as a system (it has inputs, outputs, a process, and uses feedback).

Let us use the "customer order" example given above to illustrate further the level of complexity involved. For example, taking a customer order might involve:

1. Receiving the order
2. Recording the order
3. Transmitting the completed order for further processing
4. Preparing and transmitting order corrections
5. Providing particulars to the customer regarding unit price, availability of goods, and probable shipping date

Within this very general category of a business activity, there are obviously many other factors that could affect the complexity of the problem. In other words, not all business activities are equally complex. For example: "Are orders received centrally or at many physically remote points?" "Are there special requirements involved in determining what happens to the order once it has been recorded?" These and many other questions will, when answered, affect the level of problem complexity.

Business Function Subsystem

A business function is a logically related series of activities which, taken together, represent the primary responsibility of a single business unit. Payroll preparation is an example of a business function. This business unit is generally called the Payroll Department (also Payroll Office, Payroll Unit, or Payroll Section, depending upon the size and organizational structure of the institution).

A functional subsystem is more complex than an activity subsystem in several important ways:

1. There are usually a number of different procedures involved; hence, there are more steps to be performed, more decisions to be made and a variety of possible conditions which must be considered.
2. There are usually a greater variety of input sources which must be considered; hence, there is greater interaction with outside agencies.
3. There are usually a greater variety of output requirements to be considered, which again result in interaction.
4. Because the functional business unit is a primary point of responsibility and decision-making authority, it is accountable for the accuracy and reliability of its products. This heightens

the need for internal control over the function, making for added complexities.

Looking at a more detailed picture of the payroll function this increased complexity becomes evident:

Payroll Inputs to be Considered:

1. Employee time records (hourly employees)
   a) Hours worked (regular and overtime)
   b) Jobs worked on
   c) Supplies and equipment used as a part of the job
2. Other basis for payment (employees paid on the basis of yearly salary, commissions, or some other method)
3. Employee salary schedules including input from Personnel Office establishing individual pay rates
4. New employee and employee-change data

Payroll Procedures to be Considered:

1. Preparation of payroll data for processing
2. Computation of current payroll
3. Computation of tax, social security and other deductions
4. Computation of quarterly and year-to-date payroll information
5. Maintenance of records
6. New employee and employee-change update procedures
7. Payroll adjustments and corrections
8. Control procedures which insure the integrity of the payroll process

Payroll Outputs to be Considered:

1. Paycheck preparation
2. Quarterly reports required by the government
3. Yearly reports required by the government
4. Payroll summary reports and data files which can be used by the organization in the preparation of financial statements, budgets, and other aspects of financial management
5. Data which can be used to determine the distribution of labor, supplies, and equipment costs by job function

Business Operation Subsystems

There are times when a systems problem deals with a major institutional need. Dealing with this need requires simultaneous attention to the interaction between many functional units. Thus a business operation can be defined as a process which fulfills a major institutional need involving the interaction of multiple functional units. The processing of customer orders from the receipt of the order to the shipping of goods to the customer is an example of a business operation subsystem. A brief list of some of the functions and activities involved in the order processing operation includes:

1. Order department function
   a) Order receipt
   b) Order pricing
   c) Selection of goods

2. Inventory management function
   a) Allocation of inventory
   b) Additions and deletions to inventory
   c) Transfer of goods between warehouses and manufacturing sites

3. Credit management function
   a) Credit checking
   b) Changes in credit status

4. Accounts receivables function
   a) Issuing of invoices
   b) Updating of accounts receivables file

5. Shipping and receiving function
   a) Issuing of notices to ship including packing orders
   b) Notification that goods have been shipped and/or received by the customer
   c) Notification that goods have been returned, lost or found

Although somewhat oversimplified, the above example illustrates that the level of complexity associated with the analysis of business operations constitutes a major systems undertaking. Systems investigations at this level of complexity require a major commitment by the total organization in terms of time, people, and equipment if they are to be fruitful.

Business Systems

Advances in computer technology and the science of business management have enabled us to begin focusing on the institution as a total business system. In this kind of systems investigation, it is the total institution and not a segment which is being subjected to analysis. In a sense we are seeking an answer to the question, "How should this institution approach its business and secure its goals?"

At this level of complexity we are seeking to establish:

1. A high degree of interaction and interdependence between the various business activities, functions, and operations of the institution.
2. The ability to make predictions and decisions based upon data received from all segments of the organization.
3. The ability to fully explore the potential consequences of future management decisions based on historical and forecast data.

Some of the current approaches which have been developed for dealing with problems of this complexity require conceptual understanding of management information systems and data base management systems. These concepts will be more fully discussed in chapter 9.

CHAPTER SUMMARY

Systems analysis is an important approach used in dealing with known business problems. In order to do effective systems work it is necessary to know the kinds of problems that exist, the complexity of the problems, and their relative importance to the institution.

The importance of a particular problem relative to other problems confronting the institution is usually a matter of differing opinion within the organization. Because a limited number of resources are available, it can be anticipated that there will be competition and even conflict over the allocation of these resources. At other times the sheer political nature of an organization will cause people to differ in their assessment of a problem's importance.

The analyst should anticipate conflict and attempt to structure the systems process in a way which maximizes the potential benefits of conflict. The active and continuous involvement of users throughout the systems process will help to ensure conflict resolution through the mutual education of the participants.

There are a number of frequently encountered systems problems. Knowing the kind of problem which exists helps the analyst to direct the search for information, make an objective assessment, and anticipate the technical and nontechnical complications.

By developing an estimate of problem complexity, the analyst is better prepared to: (1) schedule required people and equipment resources; (2) estimate the potential implications of the problem for other areas; (3) determine what levels of management should be involved; and (4) determine the extent of management involvement required.

The definition and classification of systems problems is a difficult and time-consuming task. However, it is crucial to the eventual development of a new system which will accurately meet the needs of people and, hence, those of the institution. The analyst should take whatever time and effort is required for a thorough job of problem definition and classification. To short-cut the commitment at this point in systems analysis will result in a less than satisfactory problem solution.

Once completed, the definition and classification of the problem should be formally presented to all parties involved in, or affected by, the systems investigation. Because problem definition and classification is so highly dependent upon data collection and analysis, a single report is made for both. A sample of the materials to be included in such a report will be provided in chapter 5.

WORD LIST

business activity—a procedure or series of steps to be followed in performing a single task

business function—a logically related series of activities which, taken together, represent the primary responsibility of a single business unit

business operation—a process which fulfills a major institutional need involving the interaction of multiple functional units

business system—a process which focuses upon an institution as a total system

problem of accuracy—systems problem regarding the correctness with which results are obtained

problem of economy—systems problem regarding how to reduce the average cost of services

problem of efficiency—systems problem reflecting a balance of concern for responsiveness, throughput, economy, accuracy, and reliability

problem of information—systems problem regarding one or more of the following: coding, storage, organization, and/or accessing of information

problem of reliability—systems problem regarding the consistency with which accurate results are obtained

problem of responsiveness—systems problem regarding the speed with which a response can be made to a request for information or other service

problem of security—systems problem which focuses on the protection of information and/or the protection of individual rights

problem of throughput—systems problem regarding the amount of start-to-finish workload which can be performed for a given period of time

QUESTIONS AND PROBLEMS

1. The Stanley Field Sprocket, Ltd. Problem:

   At Stanley Field Sprocket, Ltd., Linda Simon, a systems analyst, has been assigned to perform a systems study of the order department. This assignment resulted from the recent loss of a major customer because of "poor customer service." Some comments from Simon's report have been excerpted below. Based on the discussions in this chapter, how would you describe the various systems problems facing Stanley Field Sprocket?

   Excerpt 1 At times, the order receipt clerks (2 persons) appear barely able to keep up with the number of inquiries being made. Conversely, at times the Order Processing Unit (8 persons) appears to have little to do.

   Excerpt 2 Recently, the amount of computer "down time" has risen to 10%–15%. No satisfactory backup procedures exist to enable personnel to work with an up-to-date inventory while the computer is "down."

   Excerpt 3 The inventory printout does not indicate inventory location beyond giving the name of the warehouse.

2. The Midtown Junior College Problem:

   Doug Hartford, a systems analyst at Midtown Junior College, has been conducting a systems study of the College Housing Office. Based on direct encounters with first-year students and their parents, Hartford concluded in a report to the vice-president for student affairs that "first-year students and their parents reported that dealings with the Housing Office were frustrating. They found it particularly difficult to obtain timely and accurate information about room assignments and costs."

In this report to management, do you think Hartford has presented the problem very well? What other kinds of data might he have sought in his study? Would you have changed the emphasis? If so, how?

3. Vital Perez is Executive Vice-President of Vital Gas Exploration Company. During an interview with Perez, you learn that he is extremely disappointed with the company's systems development efforts so far. He feels that, after years of investing in data processing systems, he still has not achieved satisfactory results.

Later, in a discussion with Walter Fardown, Marketing Director, you are surprised to hear Fardown praise the systems development group as having provided outstanding service.

Since you are relatively new to the company, you can only speculate on whether Perez has had some special difficulties with the systems development group. What is your opinion on this matter?

4. Julia Guide is a systems analyst with Consolidated Grains. Throughout the early stages of a project, Julia has had difficulty getting users to meet with her. They repeatedly cancel their appointments or are interrupted during interviews. You have been asked to comment on Julia's situation.

a. What do you feel are the problems affecting Julia's project?
b. What do you recommend that she do?

5. As a systems analyst with Broadloom Rug Company, you are frequently assigned to assist distribution facilities. In one instance, you note that the distributor has difficulty shipping goods out on time. You find no pattern such as delays in the east coast shipments or delays to a particular shipper. One order simply gets shipped on time, and the next one simply gets severely delayed.

a. What kind of problem are you dealing with?
b. What kinds of further investigation do you recommend, and why?

MINICASE—CONTEXT INTERNATIONAL, LTD.

The management has agreed to use a Team Model to develop its new financial system and has asked you to participate in selecting team members. You have the following brief profiles on each of the potential team members:

Max Force, Controller for the Clean Filters Manufacturing subsidiary. Max is a busy, dynamic individual who is highly respected throughout the firm. Rumor has it that he is slated for a future presidency in one of the subsidiaries. He is a direct, take-charge individual with a short fuse for people who don't quickly grasp his ideas.

Jerry Turnkey, Senior Accountant with the Freeflow Sludge Separation assembly plant. Jerry has made considerable progress in developing the plant's accounting system by capitalizing on the ideas and suggestions of others. He views his role as that of a manager of people and systems and not as that of a charismatic leader. He is quick to say that his staff works with him and not for him. Although he is highly regarded at the plant, he is not well known throughout the firm. He does not like to jump at the first good idea that comes along.

Judy Swift, Corporate Director of Planning. Judy is perceived as one of the brightest young people on the corporate staff. Although employed at the firm for less than a year, she already has a comprehensive understanding of the business and displays an extraordinary capacity for absorbing detail without sacrificing the global view. Although nominally a "director," she is actually a one-person office with direct formal lines of responsibility to the corporate executive vice-president. Although executives within the firm know of Judy, she is viewed as a somewhat aloof loner who performs valuable staff functions.

Harold Sellout, Director of Purchasing for the Desalination Products subsidiary. Harold has worked in his current capacity for the last eight years and is perceived to have developed an outstanding purchasing system despite a lack of computerized support. While certainly not a scholar, Harold has demonstrated a great deal of common sense in getting the job done with limited resources. Harold's motto is: "Time and tenacity will get the job done." His previous experiences with attempted computerization were not particularly successful. He indicates interest in a new system but believes the approach should be an extremely cautious one.

1. Given the limited information that has been provided to you, rate the above potential team members on a scale of 1 (lowest) to 5 (highest) for each of the criteria mentioned in this chapter (that is, individual, working with others, and organizational).
2. Where is the team strongest? Weakest?
3. What kinds of conflicts might you expect between which persons if this group became the team?
4. If you could replace one person, who would it be? What would be the replacement person's characteristics?

NOTES

1. Philip C. Semprevivo, *Teams: in Information Systems Development* (New York: Yourdon Press, 1980), pp. 84–97.
2. M. E. Shaw, *Group Dynamics* (New York: McGraw-Hill, 1976), pp. 189–92.
3. E. F. Harrison, *The Managerial Decision-Making Process* (Boston: Houghton Mifflin, 1975), pp. 195–198.
4. J. R. Galbraith, *Organization Design* (Reading, Mass.: Addison-Wesley, 1977), pp. 120–122.

5

Data Collection and Analysis

CHAPTER OBJECTIVES

When you finish this chapter you will be able to—

- *State and discuss the various information levels within an organization.*
- *Describe how to approach data collection and analysis using a structured, top-down methodology.*
- *List and explain those aspects of a job which have potential significance for systems analysis.*
- *State and define nine kinds of job-related problems frequently encountered by the analyst.*
- *Construct flowcharts of a physical layout and improve the physical flow of work, given a problem situation.*
- *Construct process and paperwork flowcharts that depict the flow of work.*
- *Construct a written procedure to correct a problem situation.*
- *State and discuss three important quantitative methods frequently used in analyzing problems.*
- *Discuss documentation as a continuous process, stating how it occurs and its possible benefits to the analyst.*

In chapter 1 we briefly mentioned that one of the primary objectives of the systems analyst is to assess the job under investigation by obtaining answers to the following questions:

1. What is being done?
2. Why is it being done?
3. Who is doing it?
4. How is it being done?
5. What are the major problems involved in doing it?

Our current discussion will concern itself with collecting and analyzing that data which will enable the analyst to find satisfactory answers to these questions. During data collection and analysis the analyst is dependent upon the effective use of systems tools, such as interviews, questionnaires, standard flowcharts, HIPO diagrams, data flow diagrams, N-S charts, and decision tables, as well as other special-purpose tools which will be discussed in this chapter.

Because "raw" or unevaluated data is not particularly useful to the systems analyst, we consider data analysis as a part of the collection process. In other words, the analyst collects only data that is, upon close examination and evaluation, both relevant and useful. Thus, data collection and analysis is a screening process, not the arbitrary collection of massive amounts of raw data.

LEVELS OF INFORMATION

Because of the complexity of most business institutions, it is important to consider where to begin the search for information. Also, it is important to know where in the organization to find the "best" answers to particular kinds of questions. Different kinds of questions should be asked at different levels within an institution. The answers will provide levels of information regarding the problem under investigation. For example, the analyst should plan to collect and analyze data at the following four levels for most major systems investigations:

1. Data concerning the nature of the industry
2. Data concerning the institution
3. Data concerning management of the area under investigation
4. Data concerning actual work being performed in the area under investigation

Note that in most systems projects and particularly for those where the problem is of only limited complexity, the analyst will focus the search for information in levels 3 and 4. In the following discussions each of the above information levels will be covered.

Nature of the Industry

It is always important to know something about the nature of the industry in which an institution operates. This is true not only when the institution is a commercial corporation, but also in government and in the other vital sectors of public and private industry. Naturally, the more comprehensive the systems effort, the more crucial will be the analyst's concern in seeking answers to the questions normally directed at this level.

Some of the very vital kinds of data which can be obtained at this level include:

1. Data regarding current technology and technological trends within the industry

2. Data regarding the rate and potential impact of change which characterizes the industry

3. Data regarding the potential for growth of the industry as a whole

4. Data regarding the extent of competition within the industry

5. Data regarding the extent to which factors such as "customer service" and "public image" represent major industrial considerations

Trade journals produced by the industry, government reports, stock market reports, association conventions and publications, industry surveys, and other similar sources can supply information at this level.

By gaining a reasonably complete understanding of the industry, the analyst is better equipped to understand why certain approaches and concerns are being expressed within his or her institution.

The Institution

As we ask ourselves why things happen as they do within an institution, it becomes necessary to review that institution as a business totality. In so doing, we are able to collect and analyze data concerning the following topics:

1. The stated goals and purposes of the institution

2. The relative position of the institution within its industry

3. The "posture" of the institution in terms of the cash, personnel, and equipment resources which it has available

4. The written and unwritten policies of the institution which affect the way things get done

5. The formal organizational structure of the institution

The sources for obtaining this level of information are once again placed relatively high within the institution. Direct contact with top management is essential in many instances. Also, much of the literature produced by the institution (and therefore with management's stamp of approval), such as annual reports to stockholders, should be reviewed. In this respect formal organization charts, financial statements, policy memoranda, public relations materials, and annual company reports are important starting points for the analyst.

Understanding the institution as a whole, the analyst will be able to relate how job specifics assist or hinder the institution in meeting its stated goals and purposes.

Area Management

As we begin to look more closely at the specific problem in question, a primary source of information is the management of the problem area, or as it is sometimes called, *middle management* or *line management*. This level of management is generally considered to be the point of real authority and responsibility within the organization regarding *what* gets done and *who* does it. At this level there is also some opportunity to influence *how* things get done, but as we shall see in the next section, that influence is not always as strong as might be expected.

Because line managers are generally held accountable for what happens within their departments, their involvement in the systems process is essential. Their comments, suggestions, and approval will be important to the success of the project. In working at this level the analyst will collect and analyze data about the following (as perceived by line management):

- Area needs and demands
- Area effectiveness
- Availability and effective use of resources
- Organization of personnel and equipment
- Existing problem areas and their causes

It will also be necessary for the analyst to make an assessment of line management's responsiveness (a) to the analyst, (b) to the evaluation of the department by others, and (c) to the possibility of specific changes as a result of the systems investigation.

The primary source of information at this level is the manager. However, it is also wise to seek additional data from other middle managers whose departments may be required to interact with the problem area. The judgements and suggestions of these other managers may direct the analyst to study sensitive problem areas which would not otherwise be brought to his or her attention. However, the analyst must exercise caution in attempting to follow up on information from these outside sources.

Performing the Job

It is always interesting to see at what level within an organization the wheels really spin. Frequently it is only when the analyst talks to those who are involved in the step-by-step performance of the job that many of the problem difficulties and solution alternatives come to light. It is at this level that the analyst directly observes *how* things get done.

However, it should be also apparent that the analyst needs to first obtain the data specified at the other levels before systems work at this level can be really fruitful. The collection and analysis of data regarding the industry, institution, and area management present the framework within which job methods and procedures should be evaluated. We would not expect a corporate vice-president to explain the ins and outs of processing a customer order. Likewise, we should not expect the order clerk to tell us why greater system economy or responsiveness is required.

Some of the important data which can be collected and analyzed at this level includes:

1. The actual duties and responsibilities of personnel doing the job

2. The informal organization of personnel

3. The resources available and their use

4. The needs and demands of personnel doing the job

5. The responsiveness of these personnel (a) to the analyst, (b) to the evaluation of their work by others, and (c) to the possibility of specific changes as a result of the systems investigation

As in the case of area management, it will be essential that the analyst obtain comments, suggestions, and cooperation from these personnel in performing the investigation. They are the primary source of information at this level. Evaluation from external personnel whose jobs interact with the problem area can be viewed as a secondary information source. However, in view of our prior discussion regarding conflict within an organization, use caution in relying too heavily on secondary sources.

Up to this point, we have discussed in fairly general terms the level in an organization we look to for particular kinds of data. Once we begin to collect and analyze data about the job as it is being performed, there are certain other matters to be considered.

APPLYING A STRUCTURED, TOP-DOWN APPROACH

As mentioned earlier, the manager and workers of the functional areas being studied are the primary contact points for understanding what is being done, who is doing it, and how it is being done. In attempting to collect data on and analyze these business functions, the analyst must be careful to avoid several potential pitfalls, including:

1. Getting sidetracked into irrelevant and overly detailed data collection
2. Missing important data about processes and their relationships to one another
3. Obtaining piecemeal evidence that does not relate to a wholistic view of the problem

The analyst can avoid these pitfalls by using a structured, top-down approach. The following example shows how an analyst might approach data collection and analysis using a structured approach.

Mary Eggleston is a senior systems consultant working for the Fairview Typewriter Company. Management has determined that improvements are required in the Accounts Administration Unit, and Mary has been assigned to perform the data collection and analysis. It appears from her discussions with top management that the systems problems are not only in the area of economy, but also in the availability of management information. She wants to test these premises throughout her search.

Mary has decided to apply a structured approach using the HIPO technique. Her primary objective in an initial meeting with Bud Morgan, the office manager, is to identify what major functions are being performed by the unit and who the responsible parties involved are. Mary finds that Bud is very aware of management's keen interest and that consequently he is perhaps overly eager to talk about his problems.

Mary assures Bud that they will work closely on this study and that she will want to hear much more about his ideas. She adds that she will need his help, and that a good starting point would be for him to identify the major functions his office performs. To get him started on the right track, she sketches an overview VTOC diagram depicting what she understands to be the two major functions of his office, as illustrated in figure 5-1.

Bud agrees that these are the major functions but points out that he has not organized his office this way, because there is a high

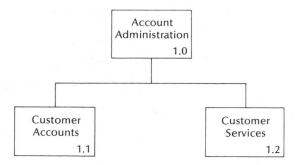

Figure 5-1 A VTOC diagram showing two major functions

degree of interaction in providing total customer service. Mary indicates that she is not yet ready to talk about how the functions are performed or how they might be improved. She is simply trying to get an overview picture of what is being done. She asks if he would help her describe the customer accounts function in a bit more detail. Through these discussions, they can expand the VTOC diagram, as illustrated in figure 5-2.

At this point, Bud is beginning to feel comfortable with the process and suggests that they focus next on the customer services function, which he completes in short order, as shown in figure 5-3. Reflecting on this VTOC diagram, Bud comments on the difficulties in trying to perform effective customer services without primary access to the latest billing, accounts receivable, and customer credit

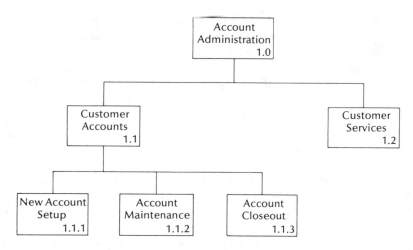

Figure 5-2 A VTOC diagram expanded from the one in figure 5-1

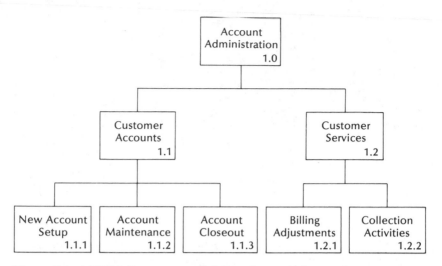

Figure 5-3   A VTOC diagram expanded from the one in figure 5-2

information. He also implies that his description of that function may not be complete. He knows that there are a lot of interfaces with other offices, and his own experience before becoming manager was in Customer Accounts. Mary obtains from Bud the name of the Customer Services supervisor and Bud's permission to spend some time speaking with that supervisor. She agrees to review the results of those discussions with Bud so that he knows what progress is being made. Mary explains that, once she has a complete VTOC diagram, she will be ready to discuss how things are being done and the problems associated with doing them.

Although the above situation is somewhat oversimplified, it does illustrate how to start applying the structured, top-down approach. Also, it underscores the necessity of forestalling the tendency of well-meaning users to drift off prematurely into discussions about "problems" they are confronting. Finally, it illustrates that, if the initial focus is on structure, the analyst will develop from the outset a wholistic and well-documented view of the problem and will be able to focus the future search for information so that it extends, clarifies, or otherwise builds on what has already been achieved.

Mary may discover that the people working for Bud Morgan have a different view of their functions than he does. If so, she may be able to assist the unit in developing an accurate and generally agreed-on statement about the functions the entire unit is performing.

Remembering the earlier discussion of structured techniques in chapter 3, we can envision that, upon completing her VTOC diagram, Mary Eggleston will begin to document the various inputs, processes, and outputs for each function—probably by using IPO diagrams, data flow diagrams, or some other structured charting technique.

However, describing inputs, processes, and outputs represents only one small aspect of Mary's work. She should not act simply as a passive recorder of what is done or how it is done. There are other matters requiring her active attention and creative insight. These matters are the subject of the next section.

WHAT TO LOOK FOR ON THE JOB

It is important for the analyst to observe and evaluate what is being done in the light of what is known about the problem. That is, if the primary problem being studied is one of responsiveness, then job performance should be evaluated according to its effect upon speed. In this way the search for information will become a screening process. If information cannot be shown to have a direct bearing on the problem and its potential solution, it is irrelevant.

It is also important for the analyst to establish the relative importance of particular observations. In this way the analyst can concentrate on factors which are most significant to the problem at hand. This quite naturally raises the question: "What factors might indicate that something of significance has been observed?"

During data collection and analysis consider the following as points of potential significance about the job:

1. What is the activity level?
2. How essential is it?
3. What other jobs does it affect?
4. How dependent is it?
5. Is a decision being made?
6. What level of responsibility is involved?

Let us take some time now to examine each of the above factors in greater detail.

Activity Levels

At times, the level of activity indicates that something is important and that it should be studied in greater depth. Therefore, it is important that the analyst determine the level of activity for each job. For example, if stored information is required to complete a job, the

High activity level for computer use

analyst may attempt to determine the number of times that information is used during a given unit of time. This could be an indication of how crucial the accessing of that information is to job completion. At other times when the focus is on a particular task, such as receiving a credit request, the analyst may count the number of credit requests per unit of time.

Activity levels of a general nature, such as "50,000 customer orders per year," are not as helpful to the analyst as those which present a more detailed picture. By recording activity levels in more frequent intervals such as a per-hour basis and by also recording these levels for different days and at different times of the month, the analyst can detect activity peaks and valleys. By actually plotting out these points of high and low activity on a graph, the analyst is able to assess the relative importance of activity levels to the total picture.

How Essential?

Although it may be true that essential jobs are also characterized by high activity, this is not always the case. To the contrary, it is very possible that highly essential jobs are notoriously low in activity. For example, a national defense alert system may display relatively low levels of activity, yet be considered quite essential. The opposite is also true: high activity levels are sometimes more a result of habit than necessity. For example, the distribution of daily computer printouts is sometimes shown to be of little real value to the people who receive them.

The task of the analyst in this respect is to determine what is essential and why. In other words, does the system require it to function properly? Or is it being done because someone thinks it is a good idea or because it has always been done this way?

What Other Jobs Does It Affect?

Many times it is the effect which one job has upon subsequent ones which makes it particularly crucial. It is important for the analyst to observe and record these kinds of inter-job relationships. For example, if the results of a particular routine job will become the basis for important management decisions, the analyst will want to attach special importance to the investigation of this job.

How Dependent Is It?

The examination of job dependence is the reverse of the examination of job effect. In other words, we are now looking at a job and asking how the performance of this job depends upon the quality of performance of those jobs which affect it. Obviously, there are times, such as in the preparation of paychecks, when those jobs which directly precede it are crucial. If timesheets are not properly filled out and carefully edited, paycheck errors will result. Noting that a particular job is highly dependent on other preceding jobs is an important key to directing the search for information.

The Decision Focus

To a large extent, understanding an organization depends upon knowing where decisions are being made and who is making them. By observing and recording the exact location and nature of these decision focal points, the analyst frequently gains insight into how things happen and why problems occur. Certainly, the fact that a decision is being made warrants some further investigation regarding:

1. The basis for making the decision
2. The potential consequences of each decision alternative
3. The criteria for "good" and "bad" decisions
4. The "track" record under the current system (for example, how many sales were lost as a result of bad decisions)

Responsibility Level

Determining just where responsibility lies within an organization is not an easy task. Title, rank, salary, and appearance of individuals are the most frequently applied criteria, but they are often misleading. Close analysis reveals that, on occasion, responsibility is maintained in the least likely places. There are wide-ranging examples in which managers insist on having a role in routine clerical decision-making, or in which clerical personnel are delegated or assume management-like responsibility.

From a purely practical point of view the analyst must understand the informal, as well as the formal, responsibility structure of the organization. The analyst must also be able to assess the association between the assignment of responsibilities and those problems which are found. In this way it may be possible to determine if there is a direct relationship between the problem and the way in which responsibility has been delegated or assumed.

RECOGNIZING APPARENT PROBLEMS

In addition to those points of potential significance previously discussed, there are a number of more or less apparent problems which may exist and to which the analyst should remain sensitive. If any of the following exist, they should be flagged by the analyst as immediate problems and should be fully documented:

1. Job duplication

2. Job overlap

3. Frequent backtracking

4. Job inconsistencies

5. Job delays

6. Poor workload distribution

7. Job inaccuracies

8. Lack of job controls

9. Lack of job instructions

The importance of recognizing these problems warrants some extended discussion for each type.

Job Duplication

Job duplication exists when two or more persons or organizational units are performing the same job or jobs. This frequently occurs when people or offices are not organized in a manner which permits them to share information. Other causes include an unwillingness between people or offices to share information, or simply not knowing that other persons or offices have information which is necessary to do the job. As a college student you may have experienced the consequences of job duplication when you received requests for the same information from several different college offices.

Detecting job duplication is most important since elimination of the problem often results in impressive cost/time savings. Centralizing job functions and/or sharing information files will generally be a part of plans to deal with job duplication.

Job Overlap

Job overlap exists when two or more persons or organizational units are duplicating one another in part, but not altogether. To illustrate this type of problem, let us take the example of how two different organizational units, namely, the personnel and payroll offices might experience such a problem.

Casa Machine Tool Company is a manufacturing company with 1500 employees. Although the company uses data processing throughout its organization, the equipment is small and control of it is highly decentralized.

The Personnel Office has central responsibility for recruiting, hiring, training, disciplining, and firing personnel, based upon recommendations from the various departments. It keeps complete personnel records for individuals and notifies Payroll when new employees are hired and when employees resign, take a leave of absence, or are dismissed.

The Payroll Office maintains a separate set of files for preparing the company payroll and necessary quarterly and year-end reports.

The problem of job overlap occurs whenever an employee's status changes. For example, if an employee's name changes or if the employee receives an increase in pay, two offices must update two sets of files with the same information. In terms of their primary functions (payroll processing and personnel administration), the two offices perform two separate and distinct jobs.

As in the case of job duplication, job overlap can be costly and time consuming to an institution. Also, there is the possibility that employees will find the necessity of providing the same information to more than one office a nuisance. Lastly, there remains the possibility that both offices will either not receive or not process the data in the same manner, leading to a discrepancy between two or more official records.

Once again it will be the focus of systems analysis to detect where job overlap exists and to develop methods, procedures, and other devices which will eliminate the need for the overlap.

Frequent Backtracking

Backtracking is a problem when it is frequently necessary to go back and recreate all of the prior steps in a process because some unexplainable error has occurred. The key to understanding and recognizing this particular kind of problem is to note that:

- Errors are detected, but often late in the process.
- It is not immediately understandable why the error occurred.
- Resolution of the problem requires starting the entire process over from the beginning.

Some of the ways in which the analyst attempts to deal with this kind of problem include:

1. Finding different points in the process and particularly at the front where errors can be detected and resolved.
2. Encouraging the user to classify and document the error when it occurs, including its probable cause.
3. Restructuring the process in a manner which establishes several *break-out* or *restart* points. Thereafter, whenever an error occurs reprocessing can begin at the prior break-out point rather than at the beginning of the process.

Job Inconsistencies

When a single method produces different results for the same problem, inconsistencies exist. Those who have attempted to "debug" a newly written computer program have a special appreciation for the fact that even a reasonably well-thought-out process can, at times, yield inconsistent results. The explanation of inconsistencies in computer programming is apt to be found in the conditions under which the method is used. For example, is the data which the method relies upon, such as the unit price of goods, always current and accurate, or is it sometimes inaccurate or out-of-date? In the latter case, there is a possibility that the second billing of a customer will be quite different from the first.

Job Delays

Job delays are said to exist when work cannot continue either because material is not available or because people are not available to do it. It is important to differentiate between the problem of job delays and that of poor distribution of work, which is discussed in the next section. While it may be true that job delays result from poor work distribution, it is not the only or even the most frequent reason.

Work may not be available if the method of delivering it from one office to another is inefficient or is not structured in a way which treats it as more important than other materials being delivered. People may not be available because they are assigned to a second job which consistently conflicts with the job in question.

Job delays are also significant since they inevitably raise questions about the dedication and competence of employees. Statements such as "people just sit on the work in that office" and "let them wait for it" are all too typical of the descriptive statements which attest to the reasons for job delays.

Finding the sources of job delay and developing the means for eliminating them are among the most challenging and potentially rewarding aspects of systems analysis.

Poor Distribution of Work

Work is poorly distributed when some people are "too busy" while others are idle, or when people are too busy sometimes and idle at other times. In the discussion of Problems of Throughput in chapter 4, we discussed workload distribution and gave a problem example. At this point our comments will be a summary of that discussion. A review of that portion of chapter 4 is suggested at this time.

In order to determine the true extent of a work distribution problem it will be necessary for the analyst to determine:

- *workload capacity*   What is the established or estimated capacity in jobs/time period for the offices or individuals in question?
- *actual workload*   What is the actual number of jobs being received per time period?
- *unused capacity*   To what extent is there more capacity than actual workload?
- *backlog*   To what extent is there more actual workload than capacity?

By comparing data between offices or job steps over a period of time such as a complete business cycle, the analyst can show where and when workload "peaks" and "valleys" occur. The relationship of these peaks and valleys to existing resources will also become evident.

Job Inaccuracies

Job inaccuracies are said to exist when close examination turns up consistently invalid results. The difference between job inconsistencies and job inaccuracies is a subtle but important one for the analyst. Perhaps the best way to describe the difference is that inaccuracies are consistently invalid results, while inconsistencies are inconsistently valid or unreliable results.

A further distinction is that inaccurate results are usually the consequence of an invalid method. But inconsistent results are usually the consequence of unaccounted-for conditions. For example, if customer orders are priced using consistently faulty computations, we have a problem of job inaccuracy. Just because results are invalid consistently, the causes of inaccuracy are generally easier to detect than those of job inconsistency.

Lack of Job Controls

A process lacks job controls when there is no way to check on the quality of the work being performed. Early in this text the self-regulating and self-correcting characteristics of systems were discussed. To establish self-correcting and self-regulating functions, the system must be able to either detect its own internal deficiencies or capture and use feedback from outside. Without effective controls, existing inaccuracies and inconsistencies will be transparent or masked.

When dealing with situations in which there is an apparent lack of control, the analyst must first determine what are the logical points in the process where controls can be effectively placed. A second consideration is a bit more complicated. It takes people and time and other resources to institute controls. Therefore, for every point of control the analyst must consider both:

- The cost of instituting controls at that point in the process
- The potential cost of errors resulting from a lack of control at that point in the process

In the ideal situation the analyst will place controls at points where the cost of control is low and the potential for costly error is high. The analyst should keep in mind that all situations are not alike. If some errors, such as invalid cost data, are not detected early in the process, they will cause many additional errors. For others, such as an inaccurate customer telephone number, it may be less expensive and equally efficient to check and correct them when they are used.

Lack of Job Instruction

A lack of job instruction exists when there are either no guidelines to tell people how to do their jobs, or when guidelines are poorly written, multiple and conflicting, or inaccurate. There are two different but equally negative consequences that can occur as a result of lack of job instruction. The more obvious one is that workers will simply not know how to do their jobs properly and work quality will suffer. The more likely consequence is that the organization will lose flexibility because it will become reliant upon particular individuals who have learned how to do specific jobs only. In this situation the organization must depend upon the availability and willingness of these individuals in order to get the job done correctly.

The problem is complicated by the fact that employees may recognize the obvious—namely, that as long as there are no written instructions they are essentially indispensable. Given this environment, detecting the lack of instructions may be easier than convincing people to document and share with others what they have learned over a considerable period—perhaps even a lifetime. Anticipate that it may take several efforts to obtain a complete set of instructions and don't assume that the first set will cover all cases. There may be a lot of situations that are "a little different," which you will learn about when new personnel try to perform the job.

METHODS FOR ANALYZING PROBLEMS

We have already seen in chapter 3 how a number of structured and traditional systems tools can be used to analyze problems. In data collection and analysis several methods are commonly used in addition to those systems tools, including:

1. Flowcharts of the physical layout
2. Process flowcharts
3. Paperwork flowcharts
4. Written procedures
5. Various quantitative or statistical methods

Physical Layout

Many problems can occur as a result of physical planning which does not lend itself to the work that must be performed. To use a very obvious example, when there is considerable physical distance between people and the resources which they require to do their jobs, delays can result. This could be the case when an order clerk

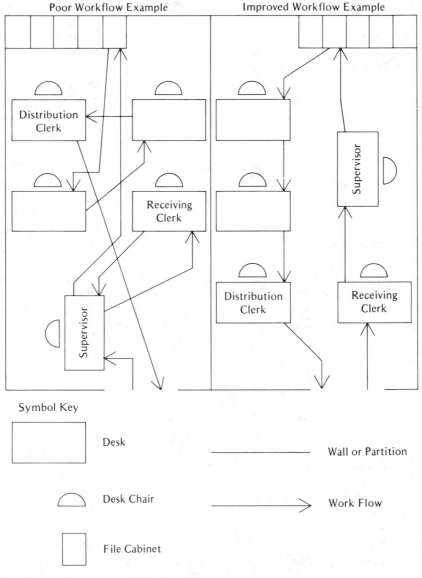

Figure 5-4 Flowcharts of the physical layout

upon the receipt of each order has to leave the desk, walk across the room to a file cabinet, jot down information, return to the desk, and complete the order form.

Using a scaled template, you can prepare a drawing of the existing layout, including the normal work flow (see figure 5-4). Then you can evaluate its efficiencies and inefficiencies, and experiment with revised layouts which might better meet work flow needs.

Process Flowcharts

Process flowcharts have been used traditionally to reconstruct and analyze manual systems and procedures. They utilize a limited number of symbols, each of which is briefly described below:

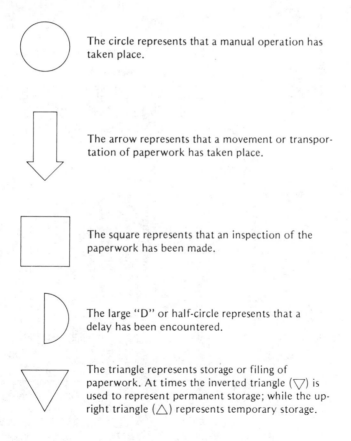

The circle represents that a manual operation has taken place.

The arrow represents that a movement or transportation of paperwork has taken place.

The square represents that an inspection of the paperwork has been made.

The large "D" or half-circle represents that a delay has been encountered.

The triangle represents storage or filing of paperwork. At times the inverted triangle ($\bigtriangledown$) is used to represent permanent storage; while the upright triangle ($\bigtriangleup$) represents temporary storage.

Figure 5-5 illustrates the use of a process flowchart. By employing process flowcharts during data collection and analysis, it is frequently possible to detect critical points in the flow of work.

Summary	Current		Proposed		Differ.	
	No.	Time	No.	Time	No.	Time
◯ Operations	5	43	3	12	2	31
⇨ Transportations	5		4		1	
☐ Inspections	1	17	1	17	0	—
D Delays	2	31	0	—	2	31
▽ Storage	2		2		0	
Distance Traveled	2050 ft.		1800 ft.		250 ft.	

Procedure Name

Admissions Application Processing

Charted By

PCS

Date: 4/14/82

Page 1 1

Step No.	Description	Oper	Trans	Insp	Delay	Store	Distance in feet	Time	Action to be Taken
1	Receive and record application	●	⇨	☐	D	▶		5	
2	Send application to Evaluation Unit	◯	⬤	☐	D	▽	300		
3	Evaluate application	◯	⇨	■	D	▽		17	
4	Application sent to Notification Unit	◯	⬤	☐	D	▽	250		eliminate
5	Update Application File	⬤	⇨	☐	⬤	▽		6	eliminate
6	Computer request prepared	⬤	⇨	☐	D	▽		2	
7	Send request to Computer Center	◯	⬤	☐	D	▽	750		
8	Computer processing	⬤	⇨	☐	D	▽		5	
9	Receive processing completion notice	◯	⬤	☐	D	▽	750		
10	Prepared individualized letter	⬤	⇨	☐	⬤	▽		25	eliminate
11	Mail letter to applicant	◯	⬤	☐	D	▶			eliminate
12		◯	⇨	☐	D	▽			

Figure 5-5 Process flowchart worksheet

Paperwork Flowcharts

Using the standard ANSI flowchart symbols discussed in chapter 3, it is possible to analyze the paperwork flow of an organization.

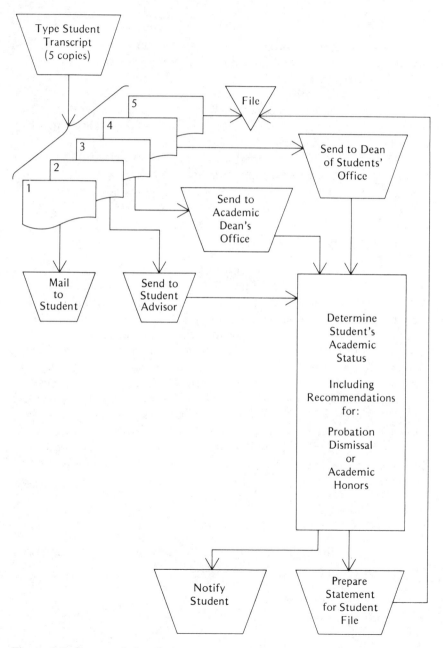

Figure 5-6 Paperwork flowchart

Figure 5-6 illustrates this method of flowcharting. A very definite advantage of this method is its use of the same standard symbols used in systems and program flowcharts.

Written Procedures

As an alternative to charting methods, written procedures are frequently employed by the analyst in analyzing problems. When written procedures are already available, the primary task of the analyst is to follow jobs through each procedural step, noting exceptions and problems as they occur. When they are not available, the analyst can construct a set of written procedures as each step of the job is being reviewed.

In preparing a written procedure there are a number of rhetorical considerations which should be kept in mind. A good procedure is:

- Clear
- Concise
- Done in an authoritative style

There are also a number of content considerations that should be considered by the analyst. In particular, a well written procedure is not simply a check list of job steps. Rather, it is a comprehensive statement about the job. For example, a written procedure should define the process, the resources, and the materials used, as well as the purpose of the process. Complete explanation of the "how," "when," and "where" of forms used should be given. A job duty statement, giving the role of each employee in the job, is also essential since it relates the job to the person performing it. Lastly, the procedure should relate the specific process in question to other processes which precede it, follow from it, or have overall management responsibility for it. Figure 5-7 illustrates in more detail a format which might be considered in actually writing a cash-sale receipt procedure.

Quantitative Methods for Analyzing Problems

There are a number of important quantitative or statistical methods for analyzing problems, including:

- Time and motion analysis
- Job sampling
- Job audits

Subject:	Date Issued	Page	of
	/ /		

Cash-Sale Receipt Procedure	Date Revised	Dept.	Code
	/ /		

  I. Definition and Purpose
   A. Define a cash-sale
   B. List all forms to be used by name and form number
   C. State the purpose of the process
 II. Explanation of Forms
   A. Indicate where and when the form is used
   B. Indicate the number of copies prepared for each form and their distribution
   C. Explain how they are to be used.
       Note: Also attach sample copies of the forms at the back of the procedure
III. Control of the Materials
   A. Who is responsible for control?
   B. What are the rules for control?
   C. How should problems be handled?
 IV. Duties of Each Employee Involved in the Primary Process
   A. General duty statement
   B. Alternative duties for each known variable condition. For example, in the case of a cash receipts counter clerk, duties should be stated for each of the following conditions:
       1. If counter clerk is not authorized to price sales or accept payments
       2. If counter clerk is not authorized to price sales, but is authorized to accept payments
       3. If counter clerk is authorized to price sales and accept payments
   C. What to do with payments and completed forms
  V. Duties and Guidelines for Related Processes
   A. Cancellation of cash-sale receipts
   B. Corrections to previously processed forms
 VI. Management Review Responsibilities
   A. Branch office management responsibilities
   B. Regional office responsibilities
   C. Home office responsibilities
   D. Internal audits and independent audits

Figure 5-7 Cash-sale receipt procedure outline

The analysis of time and motion or, as it is usually called, a time and motion study is an attempt to determine how long it takes to perform a particular task. For example if one of the steps in a particular process is to look up a customer's name, address, and account number (all of which are stored on the same index card), what is the time required for each of the associated tasks, such as going to the file, accessing the file, copying down the required data, returning to the desk, and processing the form?

By associating time to the specific act or motion, the analyst is able to generalize findings in several important ways. First, it is possible to detail where in the process time is being consumed. Based upon this it will be possible to better assess the time-saving impact of specific proposals for problem solution. Lastly, the time and motion data may provide a reasonable basis for estimating the consequences of increasing or decreasing the workloads for a given person or office.

The sheer burden of time and personnel required to thoroughly analyze each and every job transaction frequently requires that the analyst use a technique known as job sampling. In its most unsophisticated form, this means that the analyst simply selects a few job transactions for detailed analysis. In its more sophisticated varieties some scientific care is taken to insure that the job sample is truly random and, therefore, more representative of all possible transactions. Most professional analysts attempt to take at least some precautions in sampling since it is a proven fact that better sampling techniques will provide a more representative selection. While an advanced discussion of sampling methods is beyond the scope of this introductory test, the student is advised to plan for at least some advanced academic study of this important topic.

Unlike job sampling, a job audit concerns the progression of a job from the beginning to the end of a process. The analyst takes a job and follows it step by step, desk to desk, from start to finish. Using this method the analyst is better able to:

1. Obtain an overview understanding of the total process.
2. Note where in the process potential problems, complexities, and other difficulties arise.
3. Examine the process for loopholes which are not assumed but which could, in fact, happen, given some premeditated action. (For example, can I "fool" the payroll system for my personal advantage?)

Job auditing is essentially the work of management auditors. If a recent in-depth audit has been performed, much of the information required by the analyst may be available from the audit report.

However, the analyst frequently finds that either an audit has not been done recently or has been done with a different emphasis and purpose. At such times the job audit is an extremely useful systems tool.

THE PROCESS OF DOCUMENTATION

Documentation is a continuous process which starts with the beginning of the systems investigation. The analyst continues to add to this basic documentation at critical points throughout the systems process. A distinction should be made between the final information package—called Systems Documentation—and the documentation process. Systems Documentation is a final product which includes summary information about the nature of the system, how the system operates, and how people interact with the system. This final information package is a product of a three-step screening process which we call the process of documentation, and which includes:

1. Recording of relevant data
2. Organizing recorded data
3. Preparing formal reports and other documents

The recording of relevant data, as previously noted, is a selective action. That is, not all data observed by the analyst is recorded. To be recorded, data must be truly relevant to the problem-at-hand or to its eventual solution.

The organization of recorded data is an interpretive action. Using the base of recorded information, the analyst attempts to structure and interpret relevant information in a way which reveals:

- Significant new findings
- Alternative actions to be considered
- Recommendations of the analyst
- Follow-up evaluations of previous decisions

In many Systems Departments the organization of recorded data is encouraged by requiring systems analysts to prepare periodic summaries or project status sheets for each job they are assigned to. There are some advantages to such an approach. First, requiring the preparation of summary/status sheets ensures that the analyst will, at least periodically, take the time to organize and interpret data which is relevant to the problem under investigation. Second, these summary/status sheets will, over a period of time, build a basis for complete final documentation of the new system.

Preparing formal reports and other documents is the third screening action involved in effective documentation. It is, in effect, a summarization of organized relevant data to reflect the most current available information and interpretations. This third screening action results in a number of special-purpose reports and in the final Systems Documentation—each of which will be discussed shortly.

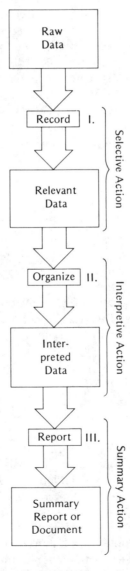

Figure 5-8 Documentation as a three-step screening process

Figure 5-8 illustrates the process of documentation as a three-step screening process. It attempts to equate recording relevant data to a selective action, organizing recorded data to an interpretive action, and reporting data to a summary action.

Presenting Results

There are a number of crucial points in the systems process where it is necessary for the analyst to prepare formal written summaries. These written summaries or special-purpose reports assist the analyst in several important ways. First, preparing a written report forces the analyst to organize and reflect upon the information to be contained within it.

Secondly, because the report is generally well conceived and well organized, it is useful as a basis for discussion with others who did not work on its preparation but are affected by its content. Finally, the approval of the report lends a kind of substance to the systems work which is yet to follow.

Let us take as an example the process of Problem Definition and Classification. Once the basic collection and analysis of data has been completed and the problem adequately defined, the analyst is ready to begin developing alternative problem solutions which can be then submitted to feasibility tests—at least it would be nice to think so. In actuality, it is likely that there will be disagreement and misunderstanding regarding the problems and their significance. Rather than proceed blindly in the hope that differences of opinion will eventually disappear, the analyst can use a special-purpose report to clarify the problem for nonsystems personnel. It will provide a focus for rational discussion of important questions and points of conflict to be resolved before proceeding further.

A typical approach to preparing a special-purpose report on problem definition and classification would be to include coverage of some of the following topics:

Background Information. It should be clearly pointed out what factors or events led to the current systems efforts. For example, if it was an audit report which led to the involvement of systems personnel, that report should be cited and pertinent excerpts included in the discussion.

Some space should also be reserved for a discussion of how the problem was initially viewed by management and the other parties involved.

The extent of the initial systems effort leading to this report should be included. The nature and extent of systems and nonsystems personnel involvement are particularly important items to be fully explained. Likewise, it is important to note which portions of the organization were studied and the names of the cooperating individuals representing various organizational units.

Through the background information section of the report, the reader will be informed of the extent of the systems effort and the variety of opinion which was considered in developing the document.

Summary Statement of the Problem.   It is important to state the problem from a systems perspective. In so doing it is usually wise to begin with an overview statement regarding the nature, complexity, and relative importance of the problem. Follow this overview with a brief summary for each aspect of the problem.

Describe important problem specifics. It may be helpful to give several examples of how certain factors contribute to the problem. In this way the reader will understand "why" there is a problem, say, of responsiveness.

The complexity of the problem is frequently best displayed graphically, using flowcharts and other diagrams. Understanding that the problem involves several different functional areas, for example, will be essential to the reader's understanding of why a major systems effort is being proposed later on in the report.

The relative importance of the problem should be presented in a way which covers all levels in the organization. The report should clarify why the problem is important to top management, to middle management, and to the person performing the job.

As part of the problem summary some general conclusions should be drawn. In particular some mention should be made regarding the level of institutional commitment and resource support which will be required to deal effectively with the problem.

Recommendations.   There are two reasons for having a separate section of the report dedicated to recommendations. First, it permits the analyst to provide some genuine direction in resolving the problem. Second, the recommendations of the analyst provide a concrete proposal to which many people can respond. While a group might have great difficulty constructing such a proposal, they can easily respond to it with effective criticism.

Some of the topics to be covered in the recommendations section might include:

. New systems alternatives to be investigated
. Limited immediate corrective actions to be taken
. Further planning action required

- Management decisions required
- A time.table of activities for the next stage in the systems process
- Resource implications of the recommendations

In effect, this section of the report will contain a detailed systems plan of action.

Similar special-purpose reports will be required at other points in the systems process and will be discussed later in the text. For example, a description of the report to be used after feasibility testing and in developing a comprehensive systems proposal is included in chapter 6.

Whenever new equipment or equipment modification is a part of the systems effort, a formal report on the evaluation and selection of equipment is essential. Chapter 10 contains a detailed description of what should be included in such a report.

Systems Documentation

As mentioned previously, Systems Documentation is one of the final products of the documentation process. It is that portion of the total documentation which describes the final system in detail. People need it to learn about the system and how it operates. Assuming the extensive use of computing equipment, some of the information which should be included in the systems documentation package includes:

- Systems Overview
- Systems/User Manual
- Computer Processing Manual
- Systems and Programming Documentation

An overview of the system should clearly outline its purposes. If it is an Accounts Receivable System, define the accounts receivable function and how the systems process operates. If there are several different business units involved in the process, then the duties and responsibilities of each unit should be clearly stated. The use of a systems flowchart is helpful in clarifying system purposes and the roles which different offices assume in the process.

The Systems/User Manual is that portion of the documentation which clarifies job responsibility, job procedures, and the various forms and instructions which have a direct bearing on how work is done. It is important to write the user manual in a way which does not presume that the user has any data processing or systems knowledge. In effect the manual should tell people in their own terminology how to operate the system. Examples of how to fill out forms, interpret error messages, and use reports should be provided in the manual wherever possible.

Just as users require a set of complete guidelines on how to use the system, so do data processing personnel. Computer Processing Manuals generally include two separate sections, one for data preparation (such as, keypunch) personnel and one for computer operations personnel.

Data preparation personnel will need to know what source documents will look like, how to key in the data, and whether or not it is necessary to key verify as well as enter the data.

Computer operators will want to know how to run the jobs, what computer programs and computer files are required, and the number and distribution of computer reports. They will also want to know whom to contact in the event of system failure.

Last, systems documentation should include a systems and programming documentation package for use by systems analysts and computer programmers. This documentation package should include complete descriptions of all files and/or data-base elements used by the system. In addition each computer program should be fully documented to include at least the following:

1. Program name and author
2. Major functions of the program
3. Description of inputs and outputs (include samples)
4. Latest computer listing of the program
5. Program flowcharts, decision tables, and structured diagrams

The Uses of Documentation

If documentation is developed on a continuous basis as we have indicated above, it can assist the systems analyst in a number of important ways. For example, it can lead to:

1. Better understanding during problem solving activities
2. Better presentations to management and nonmanagement personnel
3. Better understanding after the system has been completed (review)

Because documentation requires the organization and interpretation of relevant data, it can actually assist the analyst in problem solving. Crucial aspects of the problem tend to become more apparent as data is organized. Likewise, the implications which one variable may have for another are more evident when information is well organized.

Often it is necessary to make formal presentations to management and other interested parties. If documentation has been developed on a continuous basis, the materials needed for an accurate up-to-date presentation are readily available. Also, historical data which has been documented can be used to illustrate major difficulties which have been encountered and how they were dealt with by the analyst. This will help to illustrate the progress to others. The ability to keep others reasonably and accurately informed with little or no advanced notice can be a major benefit from good documentation practices.

A final comment regarding the value of proper documentation has to do with its value as a means of review. There are actually two primary review benefits which can be derived from continuous and comprehensive documentation. First, the documentation is of value in that it can be used:

1. As a reference for persons wishing to know about the system
2. To refresh the memory of persons who may have previously worked in the development of the system and are now being asked to make necessary systems modifications
3. To train new personnel who are either being asked to assume responsibility for the system, or are being assigned to make necessary systems modifications

There is also a second review value which can be obtained from the total documentation package. In effect, the complete documentation of a system can, if it has been developed as a continuous process, serve as a practice case in systems development. That is, because the documentation reveals the evolutionary development of a complete system, it can be used as an effective systems training device. Some of the significant points from this kind of review include:

1. An appreciation of systems analysis as a continuous process

2. A better understanding of the kinds of information required and the tools which may be used to acquire this information.

3. Insight into the variety of individuals and individual viewpoints which may be encountered

4. An observation of crucial systems issues as they relate to a specific problem area

5. Insight into the way in which decisions affecting systems development are made.

6. An example of how to fully document the systems process and an opportunity to see why it is important to do so

CHAPTER SUMMARY

Through data collection and analysis the systems analyst obtains relevant information about the problem under investigation. It is important that the analyst recognize at what level in an organization it is possible to obtain the required information and to use a structured, top-down approach to data collection and analysis.

At the most detailed level where the analyst is observing the actual job process, there are a number of potentially significant job aspects which should be noted, including job activity, essentiality, effect on other jobs, dependence on other jobs, decisions, and responsibility.

At times the analyst will observe problems which should be fully documented and flagged for further analysis. Some frequently encountered examples of these problems include: job duplication, overlap, backtracking, inconsistencies, delays, poor workload distribution, inaccuracies, lack of controls, and lack of instructions.

In addition to the primary system tools referenced in chapter 3, a number of other tools are available and particularly well suited to the data collection and analysis process. They include: (1) flowcharts of the physical layout; (2) process flowcharts; (3) paperwork flowcharts; (4) written procedures; and (5) various quantitative or statistical methods.

The process of documentation is a continuous one that begins with the start of systems analysis and is completed only with the final product. There are many aspects to documentation, from the taking of rough notes and the preparation of special-purpose interim reports, to the preparation of a final systems documentation package. Although it is time-consuming, the benefits of documentation are well worth the investment. Documentation will assist the analyst in better understanding the problem, in presenting the problem and proposed solutions to others, and in reviewing the innermost workings of the system at some later date.

WORD LIST

job auditing—a method of following and reviewing a job process from beginning to end for one or a few job transactions

job control—a method or procedure which checks on the quality of work being performed

job delay—a situation in which work cannot continue either because material is not available or people are not available to do the work

job duplication—performance of the same job by two or more persons or organizational units

job inaccuracies—consistently invalid job results

job inconsistency—different results on different occasions for the same problem using the same method

job instructions—a set of clearly stated guidelines which tell people how to do their job

job overlap—duplication of work in part by two or more persons or organizational units

job sampling—a method of analyzing a job process for one or a few job transactions.

process of documentation—a continuous screening process which begins with the start of the systems investigation and concludes with the final acceptance of the system

raw data—data which has not been evaluated and, hence, is not useful in making systems decisions

systems documentation—a finalized information package which includes summary information about the nature of the system, how the system operates, and how people interact with the system

time and motion analysis—a method of determining the time required to perform a particular job or task

workload distribution—the relationship between the capacity for workload and the amount of actual workload being assumed, as compared either at different times for the same office/individual or between different offices/individuals

QUESTIONS AND PROBLEMS

1. Below is an example of a poor physical layout. Construct a version of the layout which will result in improved workflow.

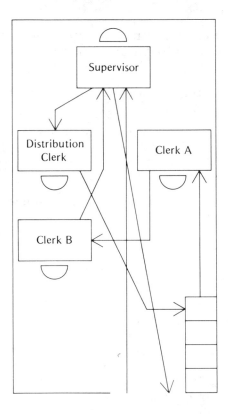

2. The Order Department manager at Rugbee Widgets Corporation has asked the Systems Department to study her operation. She feels that the time needed to accurately process orders is excessive. Furthermore, she indicates that errors the supervisor learns of are not corrected readily. You have been assigned as the project analyst.

    a. What kind of problem or problems appear evident?
    b. What kinds of help might you provide to the Order Department?

3. Construct a paperwork flowchart for the following problem: Six copies of an order invoice are prepared by the computer. Two copies are sent to the order department where one of them is

filed and the second is sent to the customer as an advance notice. The third copy is sent to the accounts receivable office where it is used to update the customer's accounts receivable record. The last three copies are sent to the warehouse. One of these copies is forwarded with the goods which are shipped to the customer. Another copy is sent by the warehouse personnel to the accounts receivable department as a notice of "goods shipped." The last copy is retained as a permanent file copy at the warehouse.

4. You have been asked to study the purchasing and accounts payable functions at the corporate headquarters of Ingot Mining Company.

    You note that, to purchase goods, one must prepare a purchase requisition and submit it to the Purchasing Department. Purchasing then checks with General Accounting to determine whether or not the requisitioner's departmental budget can cover the cost of the requested goods.

    After this approval from General Accounting, Purchasing prepares five copies of a purchase order—one copy each for General Accounting, Vendor, Accounts Payable, Ordering, and Purchasing.

    The Vendor invoice is sent to General Accounting, which then verifies that the goods have been received and issues a notice-to-pay document to Accounts Payable. Upon payment, Accounts Payable informs General Accounting and Purchasing of the payment.

    You are asked to:

    a. Diagram the above process, using your choice of structured diagrams.
    b. Comment on the specific kinds of problems you expect to encounter, based on your initial observations.
    c. Describe and diagram an alternative approach that should resolve the major problems in the above process.

5. Midway Trucking Company has been developing a large-scale system for the last nine months. Suddenly, several key members of the project team resign, including the project manager.

    You are asked to step in and keep the project on schedule. Very quickly, you learn that little or none of the completed work was documented.

    a. What do you propose to management?
    b. What do you ask the remaining staff to do immediately?
    c. What caused the problem, and how should similar problems be avoided in the future?

6. Documentation has been described as a continuous process. How does documentation become continuous? What are the potential benefits of treating documentation as a continuous process?

7. Review the situation described earlier in this chapter, in which Mary Eggleston was performing a preliminary analysis of Fairview Typewriter Company's Account Administration Unit. In her discussions with the Customer Services Supervisor, she learns that, to perform collection activities, one must:

   a. Review the current accounts file and possibly the pending new account and closed account files to retrieve customer telephone and address information.

   b. Obtain current accounts receivable information from the Accounts Receivable Department to verify that no cash has been received since the collection problem was reported.

   c. Repeat step "b," above, each time a follow-up collection action is performed.

   d. Maintain complete documentation of all collection actions taken.

   Construct an IPO diagram that illustrates how the collection activities function is performed.

MINICASE—CONTEXT INTERNATIONAL, LTD.

In the course of your work on Context International's financial system, you focus your attention on the firm's purchasing and accounts payable functions. You learn that two primary instigators can initiate the purchasing function. Either a special sales order is received for which machine parts are not normally inventoried, or regular sales orders are received, reducing the parts inventory below desired levels.

Special orders are sent to the Engineering Department directly from the Sales Department. A staff engineer reviews the order form for completeness and accuracy before placing the information on a work order document, which is sent to Purchasing. The Purchasing Department selects a vendor, assigns a purchase order number, and completes an official firm purchase order, a copy of which is retained in the file for that vendor.

Reorders do not require the staff engineer's intervention; they follow a Purchasing Department procedure similar to that used for special orders. The staffs that handle special orders and reorders are separate units within the department.

When goods are received, a receiving notice is completed by the receiving personnel and forwarded with the invoice to Purchasing. Purchasing retrieves its copy of the original purchase order and compares the goods ordered and the order cost to the goods received and the amount invoiced. Discrepancies are noted, and all documents are forwarded to the Accounts Payable Department for payment.

You are asked to describe these processes as part of a brief presentation to your project team. Your presentation should be evaluative, not only describing each process but also identifying apparent problems in the process. Be prepared to give illustrations of a "for instance" nature (such as what happens when someone asks how many widget orders are outstanding).

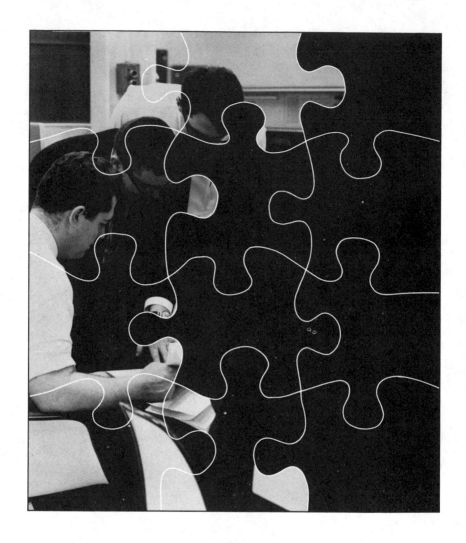

6

Systems Planning— Alternatives, Feasibility, and Proposal

CHAPTER OBJECTIVES

When you finish this chapter you will be able to—

- *Discuss the process of systems planning.*
- *Discuss and contrast the single versus alternative plan approaches to systems planning.*
- *Identify and discuss frequently encountered systems planning issues.*
- *Define systems feasibility.*
- *Discuss the effective criteria to be used in evaluating the feasibility of a proposed system.*
- *Given information about a systems problem, construct a properly formatted system plan which will deal with the problem.*

Based on the early results of problem definition and data collection, the analyst begins to consider what alternatives exist for dealing with the problem. In some instances, the analyst discovers that immediate near-term benefits can be achieved by making minor changes in the ways that people or machines are used. Often, however, minor changes simply will not do the job. In those instances, the analyst must begin planning for the major systems modifications or new systems effort that will follow. This type of systems planning should begin as soon as it becomes apparent that major changes are required and should be guided by the following considerations:

- User and management personnel should be fully involved in the process from the outset.
- An alternative planning strategy should be adopted.
- Systems feasibility testing should be the basis for selecting the final direction to be taken.
- The results should be presented as a formal systems proposal that contains clear recommendations about the kind of system to be designed and implemented.

At this point, a detailed plan is not needed, but rather a series of activities that set a more global direction for the remainder of the project. The big issues should be addressed and evaluated, such as whether a particular office should have remote terminal access to the company's computer. The more specific planning issues, such as the kind of terminal, data entry requirements, and data entry procedures will be dealt with later, during the systems design phase.

USER AND MANAGEMENT INVOLVEMENT

As with so many other aspects of the Systems Development Life Cycle, there is no one "best" time to sit down and do systems planning. However, it is neither wise nor practical to do systems planning totally in advance of an in-depth analysis of the problem. On the other hand, to wait until all of the facts are in before beginning to plan is equally unwise.

The necessity of involving management and users in the systems process has been mentioned several times in this text. Systems planning is no exception in this respect. Both groups should be involved wherever and whenever possible in the planning process. Their active participation during planning can result in greater commitment to the plan that is ultimately selected. Stated quite bluntly, by participating in systems planning, management and users acquire a "stake" in the system.

Systems planning, like so many other systems tasks, should be approached as a continuous process. That is, the analyst should begin with a very general plan based upon early problem definition and slowly begin to expand the concept of the plan or as he or she becomes more informed. At some point in time sufficient data and understanding will be available to construct sufficiently detailed plans which can be presented for feasibility testing. A graphic illustration of the time relationship between the search for information (data collection and analysis) and the planning of a system is given in figure 6-1.

During the earlier stages of the project, the great bulk of the analyst's time is spent acquiring knowledge. As the project progresses the analyst spends more and more time attempting to apply that knowledge to the development of workable systems plans. For example, during the early stages of data collection and analysis, the analyst may see certain obvious problems, such as workers hampered by the inaccessibility of needed information. While the analyst will want to give some thought to correcting this problem, to do so will be rather difficult until he or she learns more about how the needed information is stored, where it is stored, and in what form. As these additional questions are answered, the analyst, in

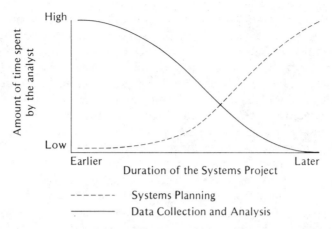

Figure 6-1 During a systems project the time spent on systems planning increases whereas the time spent on data collection and analysis decreases.

concert with management and users, will spend more time assessing the alternatives for effectively dealing with the problem. Thus, while planning begins early in the systems process, the ability to plan effectively depends upon first gaining some extended understanding of the problem.

PLANNING ALTERNATIVES

Although there are many possible ways of doing a job, frequently a single systems plan is proposed as if it were, in fact, the only legitimate way. There are many shortcomings in the approach which emphasizes a single plan of action. Three important limitations are:

1. The single-plan approach does not usually call upon the full creative capabilities of the planners.
2. The single plan becomes advertently or inadvertently identified with the people proposing it.
3. The plan might be unacceptable.

There is much talk about creativity in systems analysis, but the single-plan approach restricts rather than invites extensive creative involvement by the analyst. In effect, that approach usually means "going with" the first good idea that comes to mind. By insisting upon the creative exploration of alternative plans for dealing with the problem, the systems analyst and other planning personnel force themselves to be innovative in their approaches. Knowing that they will not be automatically required to use a particular plan simply because it has been proposed, they have the freedom to explore what appears to be formidable or unworkable on the surface. In so doing, new ideas may be developed to real advantage.

The second advantage to alternative planning is that it presents a real choice to those who will have to live with the solution. As a result, it frees the planner from the shackles of unique identity with the plan. Whenever people have been presented with a choice and with an option to participate in the selection of the "best," they also share in the responsibility for the success or failure of the plan. From a negative point of view this should make obvious sense in that it prevents scapegoating if the plan does not work. More important, from a positive point of view it enables many people to provide critical input and evaluation for all plans and to share in the good feeling that comes with doing a job well.

There are a number of frequently encountered issues which must be considered during the planning process. In this section four of these important issues will be introduced including:

1. Mechanized versus nonmechanized systems
2. Centralization versus decentralization
3. Real-time versus batch processing
4. Application versus modular systems design

Each of these issues must be decided on the basis of a particular project's requirements and not some general guidelines, if the final decision is to result in a workable plan.

Mechanized versus Nonmechanized Systems

We should not assume that to automate is always the best decision. Upon close examination, a particular problem area may be handled effectively through a redesign of the existing manual system.

In dealing with this issue the systems analyst is confronted with the fact that many people believe that one particular form of mechanization, namely, the use of computers, is the cheapest and most efficient way to do all things and that to do something manually is old fashioned. Let us take a closer look at some of the advantages and disadvantages normally associated with the use of computers:

Advantages	*Disadvantages*
1. Some decrease in the need for lower-paid clerical operations personnel.	1. Increase in equipment costs and in the cost of highly-paid technical personnel.
2. Improved processing speed per unit of workload.	2. Requires close adherence to processing schedules.
3. Access to information which either could not or would not be obtained manually.	3. Requires technical involvement (computer programming) to obtain data.
4. Less expensive when used to store, analyze, and organize business data (for example, use of payroll data in preparing required government and management reports).	4. More expensive when used to simply format and reproduce data as it is found in business (for example, paycheck preparation).
5. Machine consistency in workload processing.	5. Places some restrictions on what can and cannot be easily and quickly done.

While the above list is not intended to be exhaustive, it does show that a computer may be useful under some conditions. For example:

To assist a group of design engineers with routine mathematical calculations, a practical alternative to instant computer access is simply to give each of them a reasonably sophisticated pocket calculator.

To assist these same engineers with translating their data into detailed design requirements, computerization would be advantageous, if not necessary.

Again it is the duty of the analyst to examine this planning issue in the light of the specific problem and to weigh the relative advantages and disadvantages of each alternative in arriving at a decision.

Centralization versus Decentralization

The issue of centralization versus decentralization is a most important one with many implications for the systems analyst. At this time we will consider the issue as it relates to:

• The manner in which information is to be stored and processed
• The manner in which information is to be accessed
• The manner in which information processing and access are coordinated and controlled

	Total Centralization	Stand-Alone Computing	Star Network	Distributive Data Processing
Information storage and processing	Centralize	Decentralize	Centralize	Centralize/ decentralize
Information accessing	Centralize	Decentralize	Decentralize	Centralize/ decentralize
Coordination and control	Centralize	Decentralize	Centralize	Centralize

Figure 6-2   Centralization/decentralization alternatives

In this section, we will briefly examine four popular alternatives used to resolve the question of whether to centralize or decentralize any or all of the above systems aspects. The four alternatives are (1) total centralization, (2) stand-alone computing, (3) star networking, and (4) distributive data processing. Their features are summarized in figure 6-2.

Total Centralization

The most extreme form of centralization requires that both the information and personnel who access and use it be in the same place. For example:

Zeno Corporation is the major manufacturer of backscratchers in the United States. The greatest portion of its sales are to twenty-two distributors across the nation. Zeno has recently installed a large centrally-located computer system. In the same central location it has established an order-processing center. Using a toll-free telephone number, Zeno's major customers call directly into the order center to place orders, cancel orders, and request order information.

There are a number of traditional arguments to support this kind of centralization. For example, it is argued that centralization permits:

1. Greater inventory control through centralization

2. Greater control over the order function

3. Greater ability to share order information with other centralized business functions and operations

4. Greater economy by sharing equipment resources between the various business functions and operations

5. Greater opportunity to pool scarce EDP talent at the central location

6. Greater ability to prioritize work at a corporate level

Many feel that the development of large "super" computers for use in business validates the value of the centralized approach. They go by economy-of-scale, reflected in statements like "We can replace four of our smaller computers with a single larger one and save money."

In assessing the many alleged benefits of centralization, the analyst should be aware of two important facts. First, while benefits such as greater control, greater economy, and others have unquestionably accompanied new centralized systems, it is not clear that those results were caused by centralization per se. In other words there may have been other new noncentralized systems which would have yielded the same results simply because they, too, eliminated inefficiencies.

Secondly, it may be that arguments for centralization have concentrated only on the areas which might reasonably benefit from centralization and have not paid sufficient attention to possible negative consequence such as loss of control and flexibility at the point where business transactions are taking place.

The analyst must approach the issue of centralization versus decentralization without the preconceived notion that one alternative is clearly superior to the other in all instances. Also, the so-called "success" stories which present in glowing terms the innate superiority of one approach over the other should be evaluated cautiously.

Stand-Alone Computing

At the extreme opposite, there are times when the decentralization of both information and information access is a most reasonable alternative. For example:

Tiger Industries developed as a conglomerate during the middle to late sixties. After an initial, problem-laden attempt at centralization, the company management decided to decentralize both authority and responsibility by establishing the eighteen corporate divisions as separate profit centers. Largely because this overall movement toward decentralization was successful, many of the profit centers now permit further decentralization of their departments. For example, one profit center which is involved in the manufacturing of three different brands of breakfast cereal has two different medium-sized computer systems and three distinct order departments.

Arguments supporting the actions of Tiger Industries include:

1. Computing equipment can be selected to best meet the particular needs of the business unit being served.
2. Information and the control over that information is located at the points where business decisions must be made.
3. Accountability and profit and loss are easier to track back to the specific EDP user.
4. Step-by-step growth is allowed for as individual users are ready.

However, complete decentralization can impose important limitations on a firm's ability to obtain consolidated or integrated information about its total operations. Incompatibilities in computer hardware and software, as well as differences in the information retained and the retention formats, can occur if there is no overall coordination and control.

The Star Network

In this alternative (see figure 6-3), information is centrally stored and processed, and access to that information is decentralized. In such an environment, centralized coordination and control are necessary for security and systems backup and recovery—and this coordination and control can be difficult to achieve. Hence, this alternative necessarily involves the analyst in the whole question of data communications and the real-time versus batch processing issue, which will be discussed in the next section.

Supporters of this approach argue that the star network offers many of the advantages of both centralized and decentralized alternatives. In particular, this approach is said to let a firm:

1. Capitalize on the economy of scale associated with a highly centralized facility
2. Provide direct computer access at the location where business decisions are being made
3. Provide a well-integrated and centrally coordinated basis for obtaining information about the firm's total operations
4. Create a central pool of EDP development resources
5. Prioritize at a corporate level

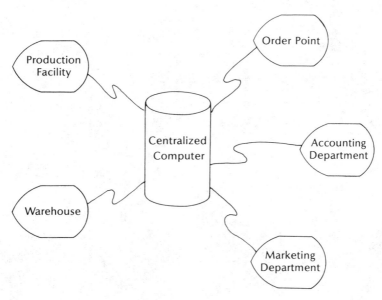

Figure 6-3   Star network

However, note that the difficulties of maintaining security increase when remote access to a system is permitted. Much-popularized stories about banking-system fraud emphasize the importance of this concern. Another concern about the star network is the entrusting of all processing to a single computer. In this environment, when the computer "goes down," no one has access to computer support. Finally, the star network introduces high communication costs, as well as potential problems of communications network reliability and availability.

Distributive Data Processing

With the introduction, broad acceptance, and use of minicomputers and microcomputers came a new concept in data processing; namely, distributive data processing (DDP). In this approach, both information and operational access to information are decentralized, but with some provision for information sharing, coordination, and control at a central level. A hierarchy of computing equipment is usually involved, as illustrated in figure 6-4. To further clarify what DDP is all about, consider the following situation.

Hearty Foods, Inc. has permitted many of its divisions to select and install their own data processing equipment. The company now has a hierarchy of seventeen different decentralized computers ranging in size from mini to medium scale. The company also has

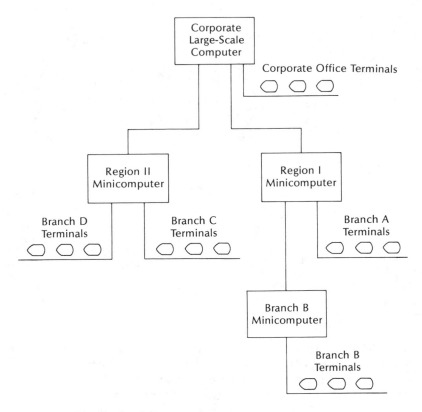

Figure 6-4   Distributive data processing

a single large-scale computer system. Three requirements have been placed on the various divisions in their selection and installation of computing equipment:

1. Computing equipment must be compatible with the company's large-scale computer in the sense that it must be capable of transmitting information to the central computer.

2. All systems and programming work must be performed according to company standards, particularly in the organization of computer files.

3. Intercomputer communication must be coordinated centrally.

By using a distributive systems approach, the Hearty Foods Company has provided the framework for both decentralized and centralized processing and access. It may also have significantly increased the total cost for equipment and personnel in the process.

The problems of effectively coordinating a distributive network are considerable and remain a major challenge of the 1980s. Notwithstanding the degree of current popularity a particular alternative may enjoy, each alternative has advantages and disadvantages. The right decision depends on problem specifics. The analyst weighs the pros and cons of each alternative as it relates to the specific systems problem.

Real-Time versus Batch Processing

Information systems have sources, processes, and uses. At times there is a physical distance between the point at which information is obtained (source) and the point at which it is processed. Similarly, there may be a distance between the processing point and the point at which the processed information is used. Whenever these conditions exist, the analyst must consider using *data communications* as a part of the systems plan. Data communications is the electrical transmission of encoded information from one physical location to another. There are a number of steps to complete the data communications process:

1. Enter input data into a business machine. This machine may be a computer terminal unit, accounting machine, or computer system that is used as an input device.

2. Encode the data for transmission. This is done by converting the information into one of several widely used communications codes. Two of the most popular coding systems are ASCII (American Standard Code for Information Interchange) and EBCDIC (Extended Binary Coded Decimal Interchange Code). The conversion of information into a communications code is generally accomplished through a device known as a data set or modem.

3. Electronically transmit or move the information. This is done via any number of different available communications lines and services, such as ordinary telephone service.

4. Decode the transmitted data. The transmitted data is converted back to a business machine form via a second modem.

5. Receive the data. This is done on a second business machine. If this business machine is a computer system, the data is generally processed and the output may be returned via the same encode-transmit-decode process.

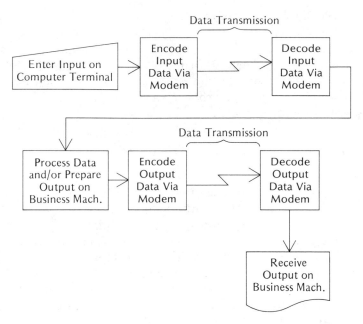

Figure 6-5 Data communications process

Figure 6-5 illustrates a full cycle of data communications from the initial encoded input information to the receipt of output for use.

At this time, rather than pursue communications technology, let us examine the impact of data communications on the analyst who is involved in planning a system. To deal with this, the analyst must consider the following questions:

• How fast should we be able to transfer information?
• How fast should we be able to access information?

While these questions may appear similar, their implications are really quite different. For example, it is possible to collect information at some remote point slowly over the course of a day, and then to transmit it at very high speeds at the end of the day. In this instance, the information is not very accessible throughout the day, even though it will be transferred at high speeds. This method of collecting information and then transmitting it in bulk for processing is called *batch processing*.

When information is being processed while business is actually taking place, we call this *real-time processing*. For example, if, while an order clerk is talking to a customer on the telephone, order

Engineer using real-time system

data is entered into and processed by a computer, this is a real-time order-processing system. In a real-time environment, we are able to both access and transfer information at very high speeds.

The issue of real-time versus batch processing is very difficult to resolve. There are benefits and liabilities associated with either approach. Matters such as need, cost, and available technical expertise must be carefully evaluated by the analyst.

Application versus Modular Systems Design

One way to plan a new system is to focus on each functional subsystem as a separate entity or application area. Using such an approach, each application area is treated as if it were totally independent; hence, there is only minimal sharing of either information

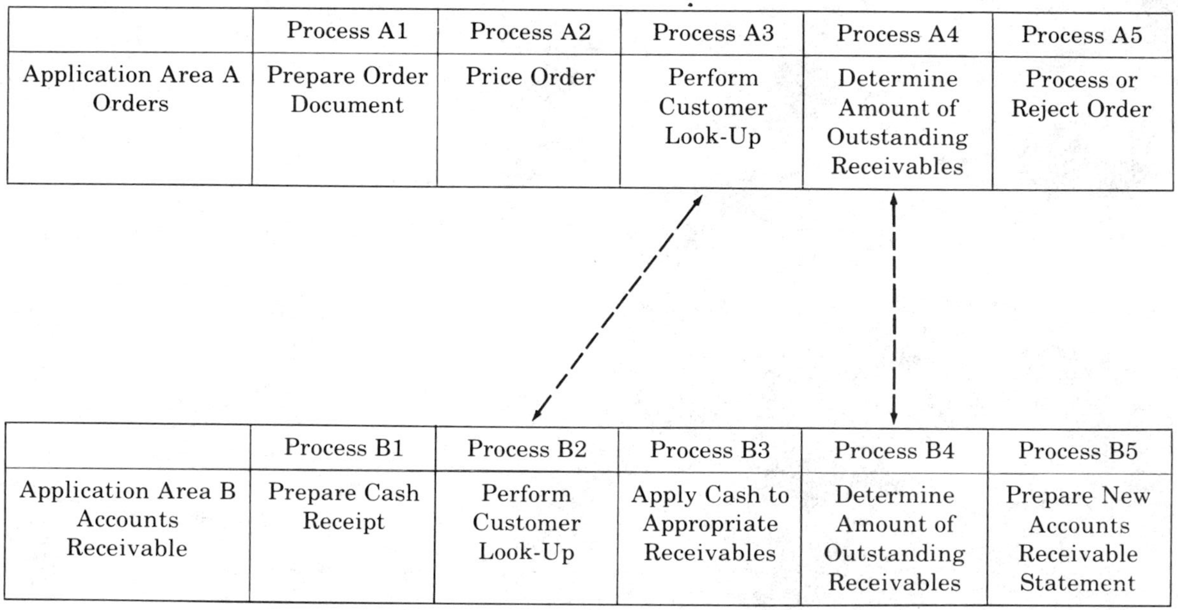

	Process A1	Process A2	Process A3	Process A4	Process A5
Application Area A Orders	Prepare Order Document	Price Order	Perform Customer Look-Up	Determine Amount of Outstanding Receivables	Process or Reject Order

	Process B1	Process B2	Process B3	Process B4	Process B5
Application Area B Accounts Receivable	Prepare Cash Receipt	Perform Customer Look-Up	Apply Cash to Appropriate Receivables	Determine Amount of Outstanding Receivables	Prepare New Accounts Receivable Statement

Figure 6-6 Application approach to systems design

or systems processes between areas. For example, if two major systems efforts were being simultaneously undertaken—say, one in the order department and a second in the accounts receivable office—these two efforts would be treated as if they had nothing in common except:

1. Orders affect the amount of receivables.
2. Amounts of receivables affect a customer's credit and help to determine if an order should be processed.

From an applications point of view, the order processing subsystem should be designed to meet accounts receivable functional requirements and vice versa. However, there would be no need to review each application area for common internal processes. For example, in each of the above subsystems, there may be a need to look up a customer name and address. Yet, each application would perform its own customer look-up and possibly do so using its own customer file. Figure 6-6 illustrates two hypothetical application areas, and how they are designed as self-sufficient entities. Note that there are two steps which are the same for each application.

The modular systems approach divides each application area into a number of smaller units called *modules*. These modules may apply to a particular application, or they may be common to two or more application areas. Modules may be used only once, or they may be used several times during the processing of an application. Special-purpose modules, which control the interaction between all other modules and their sequencing, are required for each application. Fig. 6-7 illustrates how two modularized systems might share common modules.

While modular systems design has gained many enthusiastic supporters in recent years, the decision to use it is far from automatic. Those who argue in support point out that this approach:

1. Can speed up the systems process in general, and the computer programming function in particular, by eliminating unnecessary duplications.

2. Can result in higher quality because of the concentrated effort which can be given to development of common modules.

3. Can result in greater control over the total systems project, since work can be segmented and assigned in smaller, more controllable units.

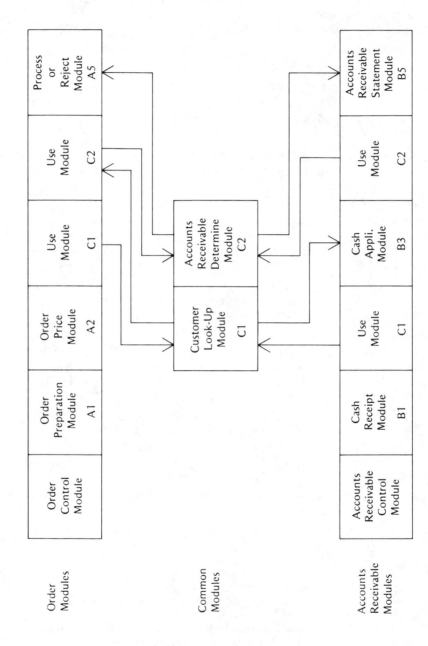

Figure 6-7 Modular approach to systems design

4. Can result in systems which are more efficiently maintained. For example, changes can be made to a common module rather than to many application areas separately.

5. Can be more flexible (open-ended), so that additional functions or applications can be added later.

Certainly these factors present a strong argument in favor of the modular approach. However, the complex realities of the business process frequently pose a number of stumbling blocks. For example, it is possible that there are:

1. Numerous unique application requirements which must be incorporated in common modules. If a single module is to accommodate all situations, it will become very large and complex.

2. Many systems changes for particular application areas. Many times a high rate of change means a high rate of potential error. When these changes and errors affect common modules, the negative consequences can be widespread—not restricted to a single application area.

Modular systems design is best viewed as one aspect of a broader planning issue, but it is not a required step in the design process. To resolve the issue of application versus modular systems design requires sound professional judgement by the analyst, based upon an in-depth understanding of problem specifics.

OTHER DESIGN CONSIDERATIONS

During the planning stage the analyst will establish the framework within which detailed design specifications can later be developed. Because much of the detailed information about systems design will be discussed in chapter 8, comments at this point will be limited to some of the more general planning considerations.

First, it is important for the analyst to review a systems design plan to insure that the system truly meets the needs of the problem area. While this may seem elementary, it frequently happens that, in the enthusiasm for creating a contemporary tentative design, an analyst loses sight of the specific problem

Second, it is necessary to review design plans as they relate to anticipated future growth. What will be the impact of increased job transactions, larger files, and other factors on the system? If the analyst doesn't allow for these matters, the proposed system might become obsolete very quickly—perhaps before it is completely installed.

Operational factors present a third important area for examination in planning the design of a system. It is crucial that the people involved be able to operate the system accurately and with ease. In reviewing this matter, the analyst should keep in mind the qualifications and capabilities of existing personnel, in order to design a system which they can and will use.

Lastly, it is important to consider how the system will be maintained once it is in use. Does the design plan adequately provide for preserving the integrity of the system through proper maintenance? Does the system provide for routine checking of results? Are errors readily detectable? Can corrections be made quickly and easily? These and other similar questions regarding systems maintenance must be satisfactorily answered by the analyst if there is to be confidence in the planning which has been done.

SYSTEMS FEASIBILITY

The term *feasibility,* like the term *systems,* is used in different ways by different people. In this text feasibility will be used to mean the test of a proposal according to some criteria. By systems feasibility we mean the test of a systems proposal according to the following criteria:

1. Ability to meet systems needs
2. Use of resources
3. Impact on the organization
4. Workability

Determining systems feasibility is, in many ways, simply the extension of the kinds of continuous evaluations that we have stressed previously in this text. The primary differences lie in the fact that systems feasibility is a test or evaluation of the complete systems plan, and not simply one aspect of it. Also, it is performed quite formally in the sense that results are presented in report form for discussion by those involved in or affected by the investigation. In other words, it becomes the basis for final selection of a particular systems plan.

Meeting Systems Needs

To establish that a proposed system is feasible, we are really asking for some proof that the system will meet stated needs before the complete system is designed and implemented. There are a number of ways of testing a proposal to establish this, and it is generally a good idea to approach the matter from several different vantage points.

First, it is possible to test the proposal logically based upon what is known about the existing system and what can be assumed about the proposed system. For example, let us assume that the problem is essentially one of throughput. There are thirteen different steps involved in each of two related processes, and some of the workload is being duplicated. The proposed system offers a reorganization of people and procedures which would eliminate all duplication. Using the actual timing of the existing processes, the analyst can estimate the overall percentage of increase in throughput which might be reasonably expected from the new system.

A second way of testing the proposal is by example. That is, locate someone who is already doing what is being proposed and examine the extent to which the system has succeeded in meeting similar needs. In using this approach, however, insure that the situations are truly comparable. If this can be established, then it is a legitimate approach.

A third approach is to perform some "live" testing of either segments of the proposed system or a limited-scale (prototype) version. This is particularly useful in testing major systems efforts.

Use of Resources

When a systems proposal is tested in terms of the use of resources, we are really asking two questions which represent opposite sides of the same coin, namely:

1. What resources does the proposed system provide?
2. What resources does the proposed system require?

The extent to which a system provides more resources than it requires is an indication of the *resource benefits* to be derived from the system. On the other hand, when more resources are required than provided, it is an indication that some level of *resource investment* will be necessary. Resource investment does not, of itself, imply something either good or bad about a proposal. This investment level has meaning only in relation to the other feasibility criteria.

Obviously the ideal situation would be to improve all other areas while realizing some direct resource benefits.

Resources are usually examined from at least the three following perspectives:

- Economic considerations
- Technical considerations
- Human considerations

The testing of a proposal in terms of its impact on economic factors is really quite straightforward. For this kind of test, the analyst demonstrates that there will be either a direct increase or decrease in the cost of people, equipment, and/or materials and supplies associated with the proposed system. A more detailed explanation of how these costs are determined can be found in the following chapter.

Testing a proposal's impact on technical factors can be somewhat more involved. It is possible that a proposed system will introduce new technical capabilities and resources of substantial value to the institution. For example, access to computers by research personnel may be a vital technical consideration because it could provide a direct resource benefit to the institution in the way of better research. In other instances, proposed systems may introduce a high level of technical complexity for which existing personnel are not ready. By examining technical factors in terms of the ability of existing personnel, the analyst can make a better assessment of the problem.

In addition to examining the relationship between proposed technology and the ability of existing personnel, some general attention should be given to the question of people as an important resource. This is perhaps the most difficult area to evaluate. What must be determined is whether or not the proposed system maximizes the use of available human resources. Does it encourage the active and creative participation of personnel? Or, does it establish rigid job requirements which ignore human needs and capabilities?

By thoroughly testing a proposed system in terms of its use of economic, technical, and human resources the analyst will have developed a second important basis for the selection or rejection of a particular proposal.

Impact on the Organization

In testing the impact that a proposed system will have on an organization, the analyst must determine if the proposed system:

1. Benefits the organization by simplifying the job process (For example, does it remove unnecessary complications in getting the job done?)
2. Clarifies the duties and responsibilities of personnel
3. Provides greater internal control over the job process
4. Has a positive motivational impact upon the total organization
5. Changes the total way of operating or fits within the existing way of operating
6. Is perceived as a benefit by the people affected

The answers are important in the overall determination of feasibility, since the effectiveness of any new system is directly related to the effectiveness of the organization which operates it. If a proposed system can be shown to have significant possibilities for improvement in this area, then its prospects for success are very real indeed.

Workability

The question of workability is basic to each of the previously mentioned feasibility criteria. If a systems proposal cannot be shown to be workable, then all of the other factors remain mere speculations. To select a particular proposal only on the recommendation of people proposing the system can be quite risky, particularly when the proposal is novel or technically complicated.

As mentioned earlier, the first hand observation of comparable systems already developed by others is one avenue for approaching the whole question of workability.

A second alternative is the use of pilot or prototype systems. This involves the actual development of a working model which can, on a limited basis, test for one of the following:

1. The workability of one or more aspects of the total system which are judged to be crucial and whose workability is open to serious question (for example, a test of a particular method of organizing and accessing information)
2. The workability of the total systems concept (for example, test of a miniature version of the system)

Certainly there are advantages and disadvantages to using the pilot approach. On the negative side, the approach requires time, people, and equipment; thus it represents some commitment of resources by the organization. It does, on the other hand, provide some real opportunity for:

- Developing a base of tangible evidence regarding systems feasibility
- Gaining insight into potential problem areas without negatively affecting the on-going operation
- Actually developing the basic concepts and techniques which will be fundamental to the design and implementation of the complete system

One point does appear essential in the decision to use the pilot approach. The proposed system should offer great potential advantages prior to pilot development. To expend valuable resources for a proposal with only limited promise would appear, at best, an unwise decision.

Feasibility Report

Based upon the results of the tests performed on the various alternative plans, a formal systems feasibility report is prepared. This report can be prepared in the following manner:

I. Problem Definition
    A. Kind of problem
    B. Complexity of problem
    C. Relative importance of problem
    D. Area(s) affected by the problem
II. Conclusions
    A. Brief description of alternatives proposed for dealing with the problem
    B. Summary of the study on systems feasibility with comparative analysis
    C. Recommendations
III. Results, advantages, and disadvantages of Individual Proposals
    A. Ability to meet stated needs
    B. Use of resources
    C. Impact on the organization
    D. Workability

This finalized report will become a major factor in the selection of one systems alternative for design and implementation.

SELECTION OF A SYSTEMS PLAN

The final selection of a system plan, while it is based upon the feasibility report, should not be limited to the analyst's assessment of it. By encouraging others who were not involved in the feasibility determination to review and critically evaluate the feasibility report, the analyst is better able to select a system for detailed design which is both sound from a systems viewpoint and has a broad base of organizational support. To obtain this support it is frequently necessary to modify a particular proposal. In this respect the analyst should attempt to:

- Remain flexible with respect to possible modification
- Fully evaluate the apparent and potential consequences of these proposed modifications
- Remain firm when he or she can demonstrate that the proposed modifications would severely limit or undermine the advantages of the proposal

Once a consensus is obtained, it is a good idea to develop in written form a finalized systems plan which establishes the basis for all further systems involvement. This formal proposal is, in effect, a contractual agreement between the systems department and the particular user area(s) involved.

It is important to select and commit to a systems plan prior to developing a detailed systems design. The development of a new, improved system requires considerable and varied resources. Up to this point the resource requirement has been essentially limited to the cost of systems, user, and management personnel for a determinable period of time. During the remainder of the project many additional resources will be required in addition to continuing the personnel commitment. Systems design requires the intense involvement of all of the parties previously mentioned plus some time from other personnel with specific knowledge about equipment, technical capabilities, forms design, and so on. Cost prohibits spreading these additional resources over a wide range of alternative plans, only one of which would be used.

Nor would a detailed design assist in resolving the planning issues, since it does not represent a workable system. That is, the design must still be programmed, tested, and implemented. Knowing what the files inputs and outputs will look like in their physical detail does not really help a great deal in settling questions about centralization, real time access, and modularity.

When a single plan is selected after feasibility testing, the analyst can concentrate the design effort and do so knowing that there is a firm commitment by the organization to a particular approach. This does not simplify the design, but it does simplify the process in that the energy and resources of the organization can be concentrated on developing a single comprehensive and detailed systems design.

Noting the logic in selecting a single systems plan for proposal, we must consider the contents and format of such a proposal. The remainder of this chapter will discuss systems proposal content and format. A concern for detail in the proposal is quite important since misunderstanding of the proposal can only help to create future difficulties.

THE SYSTEMS PROPOSAL

There is no question but that the formal requirements first for a feasibility report and then a systems proposal are time-consuming. And yet, to approach the matter informally so that word-of-mouth agreements and verbal assurances are the basis for selecting and implementing a new system is risky. People frequently forget unwritten commitments, or they change their minds regarding what is important. Through formal agreement via a systems proposal, there is some solid basis for resolving the differences which can develop between the time of the commitment and the completion of the system. The systems proposal should include a cover statement, project description, description of resource requirements, project schedule, and a list of acceptance criteria.

Cover Statement

A systems proposal should begin with a cover statement which expresses the commitment of the organization to the plan and to the project as a whole. In addition a place is usually reserved for systems and user management to formally sign the proposal.

Project Description

A description of the project and frequently a brief description of the historical factors which led to the project are provided in this section. It is also wise to include the duties and responsibilities of each office regarding its role in completing the project. If ad hoc groups, such as task forces or project evaluation and review committees, are involved in the project, then it is best to clarify their intended roles at this time.

A most important aspect of the project description section is to clearly state the intended goals of the system.

Resource Requirements

A complete description should be given regarding the resources which are required and are being committed to the project. The need for these resources in completing the project should be clearly established.

In most instances this section of the proposal will identify and contrast current systems costs to new systems costs. In addition, cost data should be related to the products and/or benefits of the new system. It may show, for example, that cost increases will result in the improved availability of needed information or that cost decreases are the direct result of the improved way in which the new system prepares a given report.

Project Schedule

It is important that the proposal spell out the time allotted for project completion, the methodology to be followed, and the nature of interim and finalized work products. In this respect, we should point out that the analyst will be better able to estimate the time required for completion of certain aspects of the job as opposed to others. For example, the analyst is in a reasonably good position to provide a systems design schedule. However, to estimate the amount of time required to program, test, and implement the system will be difficult until the design phase has been completed. Making reasonable estimates is essential and is usually the responsibility of the systems project leader. A more detailed discussion of this process is contained in chapter 11.

Acceptance Criteria

It is important to specify in the systems proposal what criteria will be used to evaluate project results. That is, it should be indicated what specific improvements are expected from the new system. For example, it may be expected that a certain minimal level of cost reduction should result from the new system.

Also, a method for reviewing systems results and for bringing the systems project to a formal close should be specified in this section. Many systems proposals clearly state that once the acceptance criteria have been met, a formal agreement or "sign-off" will be signed by all involved parties. This attests to the fact that the system has met the criteria originally set forth in the proposal.

CHAPTER SUMMARY

Systems planning is a continuous process in which systematic approaches are formulated for dealing with a systems problem. In order to be effective the systems planning process requires the active involvement of user personnel.

The development of a number of alternative plans for dealing with the problem is advised, as opposed to a single-plan approach. When presented with alternative plans, users have a real choice and the opportunity to provide critical evaluation for all proposed plans.

During the planning process consideration is frequently given to each of several major issues including: (1) mechanized versus non-mechanized systems; (2) centralization versus decentralization; (3) real time versus batch processing; and (4) application versus modular design. In addition, the systems plan should meet the immediate systems needs. The proposed system should also be within the operational and maintenance capabilities of the organization.

Systems feasibility is the test of a systems proposal according to the following criteria: (1) ability to meet systems needs; (2) use of resources; (3) impact on the organization; (4) workability. The results of the feasibility study will serve as the basis for selecting one from among the alternative systems plans for design, programming, and implementation.

A finalized systems plan is prepared as a formal, written proposal to the organization. The proposal will describe the plan in some detail. It will also describe the resource requirements, completion schedule, and acceptance criteria for the new system. It is important to secure formal agreement to the proposal by all associated parties prior to beginning the actual systems design.

WORD LIST

batch processing—a method of collecting (batching) data for processing in bulk at some later point in time

data communications—the electrical transmission of encoded data from one physical location to another

data set (modem)—a device which converts information that is encoded for business machines into a communications code which can be transmitted via a communications (such as a telephone) network and vice versa

distributive data processing (DDP)—a method of structuring data, hardware, and software into a hierarchy of processing and access that can be centrally coordinated

real-time processing—processing business transactions via a computer as they occur and obtaining the results of processing in time to effectively control the transaction

resource benefit—a condition in which a new system provides more resources than it requires

resource investment—a condition in which a new system requires more resources than it provides

systems feasibility—the test of a systems proposal according to its ability to meet systems needs, use of resources, impact on the organization, and workability

QUESTIONS AND PROBLEMS

1. Freddie Lyons, a systems analyst with Sturdy Chip Corporation, has proposed that his firm move rapidly into distributive data processing. He argues that the entire data processing industry is doing so and that Sturdy Chip may lose its reputation as a leader in data processing technology if it fails to incorporate this new technique. As manager of the Information Systems Division:

    a. What don't you like about Freddie's recommendation?
    b. How would you prefer to see these kinds of decisions made?

2. Using the descriptive example of Golden University's admissions system (presented in chapter 1), coupled with knowledge you have obtained from your reading or can obtain from your own college admissions office, prepare a hypothetical systems proposal. Be sure to devote at least some attention to the following:

Cover statement	Project schedule
Project description	Acceptance criteria
Resource requirement	Summary comments

3. Your department manager frequently relies on you, a senior systems analyst, to brief newly hired junior staff members. Today she has asked you to explain to them the key aspects of writing a feasibility report. Prepare your discussion outline and develop some illustrations to clarify major points.

4. The payroll and personnel offices perform these functions when appointing an individual to a new position:

Personnel
- Notify candidate of offer
- Obtain candidate's acceptance
- Build employee personal data record
- Notify Payroll Department

Payroll
- Obtain notice from Personnel Department
- Build employee personal data record
- Build employee salary record
- Build employee payroll record

Based on the limited information provided above:

a. Draw a VTOC diagram for each independent application.
b. Draw a single VTOC diagram for both applications that illustrates how a personnel/payroll system might be modularized.

5. Draw a diagram that illustrates a distributed data processing network in which:

a. The corporate headquarters has a large computer installation serving 350 online terminal users (it is not necessary to draw 350 individual terminal symbols).
b. Three large minicomputers are linked to a single mainframe.
c. Two of these large minis serve more than 60 online terminals each.
d. The other large mini is linked to two small minis, each of which serves more than 30 online terminals.

MINICASE—CONTEXT INTERNATIONAL, LTD.

Below is a listing of eight CIL subsidiaries and the number of employees currently on their payrolls. Assume for a moment a direct relationship between the number of employees and the amount of computing power required to support the operation of the new financial system. Thus a subsidiary with 10,000 employees would require twice the computing power of a subsidiary with 5000 employees:

CIL Subsidiary	Employees
Clean Filters Group	15,000
Freeflow Sludge Separation Group	5,000
Desalination Products Group	5,000
Aeration Products Group	5,000
Fishbowl and Consumer Products Group	20,000
Fluid Coolants Group	8,000
Nuclear Waste Treatment Group	3,500
Acid Rain Detoxification Group	4,000

You have been asked to explore the distributive data processing alternative for meeting the firm's computing requirements. The CIL approach assumes that corporate headquarters will retain its large-scale computer, which all other computers will ultimately be linked to.

In constructing your DDP network, you may select from three minicomputer levels: small (suitable for 10,000 employees), medium (suitable for 20,000 employees), and large (suitable for 30,000 employees). Or you may simply specify a bank of computer terminals by using a manual input symbol. Each bank of terminals serves 1000 employees and reduces the unused computer potential by that amount.

Finally, you should assume that a connection between two computers reduces the available capacity by 10 percent of its original amount because intersystem communications and coordination require computing power from each interconnected machine. Thus, if a computer with the capacity to serve 10,000 employees is connected to two other computers, its net available capacity is 8000 employees.

Construct your DDP network in a form ready for presentation to the class.

7

System Cost Determination

CHAPTER OBJECTIVES

When you finish this chapter you will be able to—

- *Define and discuss the kinds of costs and benefits that may exist for any system.*
- *List and discuss the cost categories which are used in determining system cost.*
- *Construct various cost tables and charts which uniformly present comparative cost data.*
- *Determine the break-even point for a new system alternative.*
- *Determine the payback period for a new system alternative.*
- *Determine the cost of data processing services for keypunching, computer programming, and computing.*

In the discussion of systems feasibility in chapter 6, we pointed out that the analyst must test a new systems proposal for its economic impact on an organization. However, there is considerable variation in the requirements placed upon the analyst in developing economic or cost data. Different types of institutions will vary in the kinds of costs that interest them and in the methods of presentation that they prefer. For example, a large publicly funded institution may not be interested in "return on investment." Private business corporations, on the other hand, may be extremely interested in this data.

Also, the manner in which funds are initially allocated or budgeted will determine in large part the manner in which cost data is presented. For example, if computer equipment rentals are allocated to one budget category and computer center personnel to another, systems costs data should be presented in a manner which respects these budgetary differences.

The interpretation of cost data is relevant to the particular situation. If the systems problem is a critical one, it is likely that management will look at the cost data to determine if it is "reasonable" or "too high." If the problem is not critical, there is a greater probability that there will be an interest in comparing the cost of the project to the cost of other projects and in scrutinizing and comparing the benefits of each.

The availability of economic resources is another factor which will vary greatly from one organization to another. It will have a dramatic influence upon what kinds of cost data are required and how that data is interpreted by the organization. When resources are scarce, systems plans which promise to increase the availability of resources or provide benefits with little or no cost increase are frequently supported by the organization.

A final but most important difference between organizations is the availability of essential cost data. The analyst should be prepared for the fact that, in some situations, basic cost data will not be available. The reason for this is simply that in some organizations the various offices are not required by management to determine detailed operational costs, such as the cost of preparing an order or a payroll check. Therefore, in many instances, the analyst will have to construct a reasonable basis for determining the per unit cost before he or she can begin to make systems cost determinations and comparisons.

To summarize, the analyst must consider the matter of systems costs within the context of organizational and situational uniqueness. The cost data, its presentation, and its interpretation will vary according to the specific organization. Furthermore, the analyst can expect a wide range of variation in the availability of required cost data.

This chapter will discuss systems costs generally and thus cover the broader range of costs which may be of interest to organizations. Because it is common in systems analysis to discuss new systems costs within the context of the benefits that the new system will provide, these issues are discussed together in the following section. Later sections of the chapter refer more specifically to methods of developing and presenting systems cost data. Lastly, data processing costs and the data processing cost center concept are discussed as related issues of special importance to the broader topic of systems costs.

SYSTEMS COSTS AND SYSTEMS BENEFITS

Systems costs are those which contribute to the cost of developing, implementing, and operating a system. Systems benefits are those products or improvements which result from the system and which benefit the organization. There are several different types of systems costs and benefits which the analyst is concerned with.

- Tangible and intangible costs/benefits
- Direct and indirect costs/benefits
- Recurring and nonrecurring costs/benefits

Because each is essential to the development of a comprehensive statement of systems cost/benefit, some separate discussion of each is warranted at this time.

Tangible and Intangible Costs/Benefits

A tangible cost is one for which the actual dollar amount is known or can be estimated. For example, a particular system requires the constant use of four computer terminals, each of which costs $100/month. The tangible cost of the terminals is:

$$4 \text{ (terminals)} \times \$100/\text{mo} \times 12 \text{ mos.} = \$4,800/\text{yr}$$

An intangible cost is one which is assumed to exist. The amount is estimated but cannot be proven. There is an important distinction between an unknown cost and an intangible cost. Frequently, unknown costs are inappropriately called intangible. If the cost can be identified as actual or provable and its amount estimated, then it is tangible. If, on the other hand, it is assumed that the system may have discouraged something (such as a customer order), then we are talking about an intangible. The estimate for a tangible cost can usually be made with considerable accuracy. The estimate for an intangible cost is an assumption. To clarify this distinction it may be well to use the following example:

Estimated cost of customer orders received but lost or misplaced (tangible)	= $30,000
Estimated cost of potential customer orders which were not secured due to poor customer image of the company (intangible)	= 24,000
Total estimated lost orders	= $54,000

In dealing with systems costs, most are tangible, and it is unlikely that systems decisions would be made solely on intangibles. When money is a factor people, and particularly managers, prefer to talk about the tangible.

Systems benefits, like systems costs, may be discussed in terms of the tangible and the intangible. However, the problem is somewhat complicated by the fact that many important systems benefits are intangibles. For example, a new terminal-entry order-processing system may be said to provide better "customer service" because the order clerk can now quickly answer the customer's question about the status of an order. What will this mean in terms of benefits?

1. Will the customer order more goods because of better service?
2. Will the customer call fewer times, thus reducing the workload in the order department?
3. Will other customers learn of this improved service and bring their business to the company?
4. Will the improved system result in higher employee morale and productiveness?

These are intangibles and it will be necessary to try and estimate the extent to which they will occur and the consequences. Obviously, the analyst cannot make a "blue sky" estimate of intangible benefits. In the above instance, marketing or sales personnel would work with the analyst in developing probable benefit outcomes.

The problem of evaluating intangible benefits would be simplified if there were many tangible benefits also associated with the system. For example, if the above order processing system also resulted in a smaller staff, lower cost per business transaction, and other tangible benefits, the analyst could place less emphasis upon intangibles. However, in many instances this is simply not the case.

Direct and Indirect Systems Costs/Benefits

Because most business systems are really subsystems within a larger entity, they cannot help but affect and be affected by other subsystems within the organization. Direct systems costs are those directly related to the expense of operating the subsystem in question. Indirect costs come from operating other subsystems which support the one in question.

For example, in order for a college admissions office to process an application for admission, there are a number of direct admissions costs. These begin with the effort to recruit applicants and are frequently not completed until the applicants actually register for classes. However, there may be a number of indirect costs to other offices also associated with the admissions process. The various academic departments at the college may, for example, make faculty available to talk to prospective students about the college programs. These kinds of supportive indirect costs are also part of the total systems cost and should not be overlooked by the analyst.

The same distinction, direct vs. indirect, may be made regarding benefits. To continue the college admissions example, improvements in the admissions system may provide support to other nonadmission functions. For example, knowing how many students are coming and in what curricula will help the Registrar's Office to develop a schedule of courses to be offered. As with indirect costs it is important for the analyst to enumerate the indirect as well as the direct benefits of the system.

Recurring and Nonrecurring Costs/Benefits

It is important for the analyst to differentiate between continuing or recurring costs and those which are one-time or non-recurring costs. For example, the cost of developing a new system is a one-time, nonrecurring cost. Maintaining and operating a new system will be a recurring or continuous cost. Similar differences exist between equipment that is purchased versus that which is rented, and one-time, start-up costs versus operating costs. In most instances there is a considerable difference between first-year and subsequent-year systems costs, and this is largely due to the many nonrecurring costs associated with the development and implementation of a new system.

If at all possible it is also wise to project the period of time over which the cost data may be accurate. It is a fact that any new system will have a finite life span. By projecting that life span, the analyst can determine when additional nonrecurring costs will be necessary. This is particularly important since it presents to management data which can be useful for long-range systems planning.

Because a system is designed to deal with continuous, on-going problems, the beneficial results of that system are likely to be of a recurring nature. There are exceptions to this rule, however. For example, if the consequence of a new system were to result in an immediate, one-time tax credit for a corporation, this action might be viewed as a nonrecurring benefit.

COST CATEGORIES

In order to determine system costs, the analyst must obtain information regarding:

- Personnel costs
- Equipment costs
- Supplies and expense costs
- Overhead costs

Building upon the earlier discussions in this chapter, we find that there will necessarily be a number of cost subcategories associated with each of the above. For example, it will be necessary to know

which personnel costs are recurring versus nonrecurring and which
are direct versus indirect. Since the analyst is primarily concerned
with tangible costs in new systems projects, the topic of intangible
costs will be eliminated, thus simplifying some of the following illustrations. However, in practice, intangible as well as tangible cost
data must be presented whenever it exists.

Personnel Costs

The cost of personnel has two parts: the salary of personnel and the
additional costs known as "employee benefits." Employee benefits
include items such as paid vacation, sick leave, retirement contributions, unemployment insurance, and health insurance contributions, which are provided by the organization. Since these
employee benefits frequently amount to 18–25 percent of salaries in
additional cost to the organization, they must be added to the basic
salary of personnel:

Salary Cost	$100,000
Employee Benefits (20%)	20,000
Total Personnel Cost	$120,000

When figuring systems costs, recurring systems personnel costs
are those which are expected to continue throughout the life of a
system. Nonrecurring personnel costs are one-time costs frequently
experienced during new systems development. For example, one-
time consultant fees, systems analysis and design costs, computer
programming costs, systems testing, employee training, and systems implementation costs are generally treated as nonrecurring
personnel costs. Those associated with the maintenance, operation,
and supervision of the system are recurring personnel costs. Figure
7-1 illustrates how tangible personnel costs might be presented by
the analyst for a new system. Only the first three years of the system are shown, for purposes of simplicity.

In this example, the impact of nonrecurring costs on the total systems cost, while considerable, is generally limited to the first few
years of the system.

Equipment Costs

Determining the cost of equipment is frequently complicated by the
fact that equipment may be shared with other systems. For example, a single computer may be used by several different offices. Thus
some fair share of computer services must be assigned to the particular system in the form of an equipment cost.

	Year 1 Tangible		Year 2 Tangible		Year 3 Tangible	
	Direct	Indirect	Direct	Indirect	Direct	Indirect
Nonrecurring Personnel Costs						
Consultant Fees	10,000					
Systems Analysis and Design	40,000	15,000	2,000	500		
Computer Programming	15,000	2,000	1,500	1,000		
Systems Testing	1,500	700	500	—		
Training and Implementation	3,700	1,000	2,000	1,000		
Nonrecurring Totals	70,200	18,700	6,000	2,500		
Recurring Personnel Costs						
Systems Operation			15,000	3,000	15,500	3,200
Supervision and Control			3,000	1,500	3,000	1,500
Systems Maintenance			1,500	750	1,400	700
Recurring Totals			19,500	5,250	19,900	5,400

Personnel Costs—Summary		Year 1	Year 2	Year 3
Nonrecurring	Direct	70,200	6,000	—
	Indirect	18,700	2,500	—
Recurring	Direct	—	19,500	19,900
	Indirect	—	5,250	5,400
Totals		88,900	33,250	25,300

Figure 7-1 Tangible personnel costs for a new system (first three years)

The problem is further complicated because what constitutes systems equipment cost (for example, five hours of computer time per week) may be much more than simply equipment cost. An hour of computer time is frequently costed to include all of the people, supplies, and equipment dollars which enable the computer to operate. Because data processing costs are a major factor in systems costs, a more complete discussion is included later in this chapter.

In presenting systems equipment costs as a part of a systems proposal, the analyst may follow a method similar to that used for personnel costs, using a variety of equipment cost subcategories:

Nonrecurring Equipment Costs
- Program development and debugging equipment cost
- Systems testing equipment cost
- Systems training and implementation equipment cost
- Equipment purchases

Recurring Equipment Costs
- Systems operation equipment costs
- Systems maintenance equipment costs

Systems Supplies and Expense Costs

Just as it requires people and equipment to develop and operate a new system, so it requires a wide range of supplies, materials, and services to support these efforts. Travel by systems personnel, training manuals, operational user guides, and other paper-and-pencil costs are a few of the nonrecurring costs typically required in new systems development. Similar kinds of supplies and expense costs may be recurring throughout the life of a system.

Systems Overhead Costs

There is a great variety of opinion among organizations regarding the extent to which overhead costs are considered essential to the development of systems cost estimates. Generally, there are two categories of overhead costs, and each may be treated quite differently by an organization:

1. Administrative overhead—some fair share of the overall cost of higher level management and/or administration of the organization
2. Facilities overhead—the cost of heating, lighting, floor space, furniture, and similar facilities used by the offices involved

Some percentage of cost for administrative overhead in a systems proposal is by far the most frequently encountered of these two items, and even this is not particularly common. The cost of facilities overhead is not generally included unless the proposal recommends a major change in facility outlays. For example, if a proposal recommended a 30 percent increase in the amount of office space to be assigned to the accounts payable department, some facilities overhead cost figure would probably be included in the proposal.

COMPARATIVE COST ANALYSIS

Comparative cost analysis is generally used to determine the economic advantages of the new versus the current system, or to determine the economic advantages of one new systems alternative to another.

Using the four basic cost categories previously discussed, figure 7-2 shows the costs for both the current system and the alternative over the anticipated life of the new system (5 years).

Current Systems Costs

	Current Year	+1	+2	+3	+4	+5
Personnel costs	15,000	15,000	19,000	20,000	23,000	26,000
Equipment costs	10,000	10,000	10,500	14,500	16,000	18,000
Supplies and expenses	5,000	5,000	5,500	5,500	6,000	6,000
Overhead costs	—	—	—	—	—	—
Total costs	30,000	30,000	35,000	40,000	45,000	50,000

New System Alternative X

Nonrecurring	Current Year	+1	+2	+3	+4	+5
Personnel costs	30,000	4,500	1,000			
Equipment costs	12,000	1,000				
Supplies and expenses	6,000	500				
Overhead costs	—					
Total nonrecurring	48,000	6,000	1,000	—	—	—

Recurring						
Personnel costs		15,000	15,000	15,000	16,000	16,000
Equipment costs		10,000	10,000	10,000	10,000	10,000
Supplies and expenses		2,000	2,000	2,000	2,000	2,000
Overhead costs		—	—	—	—	—
Total recurring	—	27,000	27,000	27,000	28,000	28,000

Figure 7-2   Comparison of costs for current and new systems

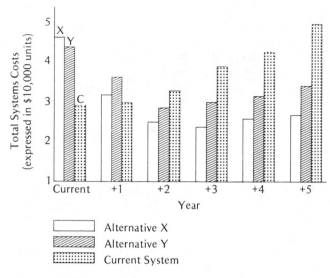

Figure 7-3 Bar-graph comparison of systems

It is at times preferable to present comparative cost data in graph form. Figure 7-3 illustrates a bar graph comparison between the current system and two new systems alternatives. By comparing these costs, the analyst can determine the following about the systems:

1. The extent of relative advantage or disadvantage for each year over the life of the system
2. Net relative economic advantage or disadvantage over the life of the system
3. The point at which relative economic advantage will begin to occur, if at all

To assist in the interpretation of comparative cost data, a number of common quantitative methods are frequently employed by the analyst, including:

> Break-even analysis
> Payback-period analysis

Some brief comment will be made regarding each of these approaches as they apply to comparative systems costs.

Break-Even Analysis

Applying the concept of break-even analysis to a comparison between two systems is somewhat different from its traditional use in

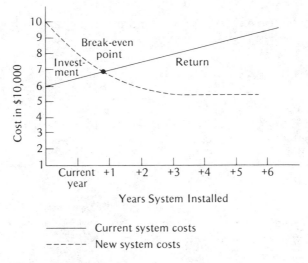

Figure 7-4   Break-even analysis chart

accounting. In accounting one compares the cost of a project to the revenues which result from that project. When the revenues equal the cost, the project is at the break-even point.

In systems analysis break-even analysis is used to compare the costs of a new system to the current system and to determine the point at which the new system will cost the same as the old. Figure 7-4 graphs the break-even analysis.

In figure 7-4, we see that while the cost of a new system would initially be higher than the current system, by the end of the next year the system would have reached the break-even point and thereafter become more economical to operate than the current system. The period in which the new system initially has a higher cost is frequently called the *investment period*. The time during which the new system has a lower cost is called the *period of return*.

Payback Period

While break-even analysis indicates when the costs for current and new system will be equal, it does not tell when the initial new systems investment will be fully recovered as a result of its greater economy. In order to do this we would have to compare the cost difference to the left of the break-even point (investment) to the cost difference to the right of the break-even point (return).

One way to do this is to chart out the annual costs for both the current and the new system, determine if there is an investment or a return on the new system, and maintain a record of how much remains to be paid back. For example:

	Current Year	+1	+2	+3
New System	90,000	70,000	60,000	60,000
Current System	60,000	70,000	70,000	90,000
(−) Invest/Return (+)	−30,000	0	+10,000	+30,000
Payback Due	−30,000	−30,000	−20,000	0

In the above example, while the break-even point will be realized in the second year it will not be until the end of the fourth year that the initial investment can be fully recovered.

DATA PROCESSING COSTS

As one aspect of making a systems cost estimate, the analyst must determine the extent to which a system requires keypunch, systems, programming, and computer resources. When the topic is the current system, this information is best obtained from records which a data processing center may keep of its services. For a new system, estimates of these requirements are frequently contained in the new systems proposal.

Next, the analyst translates these resource requirements into cost estimates. It is insufficient simply to state that a new system will require five hours of computer time per week to operate or that sixty hours of computer programming will be required to implement a new system. Rather, a detailed cost estimate is required.

The following discussion contains a reasonable approach which can be used in arriving at data processing cost estimates. The discussion is simplified in that it deals only with tangible costs and does not consider the very real possibility that data processing costs, like all systems costs, may be either recurring or nonrecurring, direct or indirect. The discussion is complicated in that determination of data processing costs is closely tied to the manner in which records of data processing use are organized and maintained.

To effectively determine data processing costs, the systems analyst must obtain information for each of the cost categories.

- Equipment costs
- Personnel costs

- Supplies and expense costs
- Overhead costs

For the purposes of this discussion the topic of overhead costs will not be treated in detail because of its limited use outside of data processing service bureaus, which are in the business of data-processing-for-profit.

Data Processing Equipment Costs

Data processing equipment costs can be broken down into two major cost categories, namely, *hardware* or machine costs and *software* or machine support costs. The first step in determining equipment cost is to identify the specific machines and machine-support packages which contribute to total equipment costs. This is accomplished by developing a complete inventory list of existing or proposed data processing hardware and software.

Determining the actual or estimated cost for processing a particular job on data processing equipment requires that the analyst also obtain information regarding the utilization of the equipment. Knowing how much the equipment is used and how much of the total use can be assigned to a particular job, the systems analyst assigns a portion of the total equipment cost to the particular job. The means whereby the analyst arrives at these equipment costs is somewhat involved. A separate detailed discussion of how one arrives at equipment costs can be found in appendix B of the text.

Data Processing Personnel Costs

In addition to equipment costs, personnel costs are important. Indeed, sometimes the major costs associated with a data processing service are personnel costs. This is generally the case when the services provided are computer programming and systems analysis. At other times, as in keypunching, there will be an almost one-to-one relationship between people time and machine time. In still a third possibility, many people are required to operate a single machine that may be working on more than one job at a time. For example, a large-scale multi-programming computer may be staffed with three full-time computer operators for each of two eight-hour shifts. Let us consider how information about personnel costs might be captured for each of the above possibilities.

It is possible to assign the work time of personnel directly to a customer job providing each customer is identified by a unique cus-

Employee Name:	M. Rushe			Supervisor's Signature:				

Employee Name: ___M. Rushe___   Supervisor's Signature:

Employee Number: _005-32-5667_   _Jo. Doodle_

Payroll Period Ending: _4/14/82_   (1) Time should be recorded in units of quarter hours

Customer Charge Number	Job Code	Mon.	Tues.	Wed.	Thurs.	Fri.	Sat.	Sun.	Week Total
44024	01	4.00	2.00	1.25		.75			8.00
52607	01		3.25		6.00				9.25
72134	02	2.50		3.75	2.00				8.25
47602	02	1.50	1.75			7.00			10.25
99999	—		1.00	3.00		.25			4.25
Totals		8.00	8.00	8.00	8.00	8.00			40.00

Figure 7-5 Modified time card

tomer charge number. For clerical and machine operating personnel, this is frequently done through a modified time-sheet or time-card system. Figure 7-5 illustrates how the time-sheet record might look for a keypunch operator. Using this data to develop workable cost data imposes some additional considerations.

The many indirect costs associated with an employee's work time, such as vacation leave, holidays, sick leave, and break-time, must be included in the per hour charge to customers. The net effect of these various employee benefits is to reduce the average number of hours per week that an employee is available to work. Therefore, it is likely that the per hour rate which a customer pays for personnel will be considerably higher (18 to 25 percent) than the per hour rate arrived at by simply dividing the weekly salary by, say, forty hours.

Professional employees such as computer programmers and systems analysts are generally paid on the basis of an annual salary or wage. Also, they are frequently not required to punch in and out of work. And yet, for an accurate account of their time, some internal record of how long they worked and on which customer's jobs is essential. Figure 7-6 illustrates a monthly summary form which might be used to record this kind of information. Because pro-

OFFICE OF COMPUTER SERVICES
SYSTEMS & PROGRAMMING APPLICATION ANALYSIS
MONTH OF   MAY '82

DEC '81-MAY '82

APPLICATION CHARGE CODE	G. WALDORF HOURS	PERCENT	K. JOHNSON HOURS	PERCENT	A. KOWALSKI HOURS	PERCENT	MONTHLY TOTAL HOURS	PERCENT	YTD TOTALS HOURS	PERCENT
86501	4.5	2.9			41.5	26.9	46.0	9.9	4462.0	17.8
86765	18.0	11.4			43.0	27.9	61.0	13.1	423.0	17.1
86555	20.5	13.0	148.5	96.4	10.0	6.5	179.0	38.4	1062.0	42.4
44024	1.0	.6					1.0	.2	10	.1
86769									0	—
44002	15.0	9.5					15.0	3.2	15.0	.6
87950	2.0	1.3					2.0	.4	57.5	2.3
86738					58.5	38.0	58.5	12.6	123.5	5.0
86835	19.5	12.4					19.5	4.2	19.5	.8
DE405	9.0	5.7			1.0	.7	10.0	2.1	100	.4
FSA									100	.4
MISC	68.0	43.2	3.5	2.3			71.5	15.4	3165.0	12.8
DCM			2.0	1.3			2.0	.4	10.0	.4
TOTALS	157.5	100.0	154.0	100.0	154.0	100.0	465.5	99.9	24800	100.1

Figure 7-6 Monthly summary of systems analysts' time

grammers and analysts usually work for extended periods of time on one or two important projects, the recording of this data should not impose any significant burden upon them. The importance of the data, on the other hand, to our costing scheme, is quite obvious.

The considerations for leave time and other variables that apply to clerical and machine-operating personnel also apply to the computation of an hourly rate for professional data processing personnel with one additional consideration: data processing professionals frequently work more than forty hours per week and without overtime compensation. At times this factor will more than negate the impact of "time off" on the billable hourly rate of work.

In attempting to calculate personnel costs one scheme should be used for computer operators and a second for all other data processing personnel. For everyone but computer operator personnel, the productive use of personnel based on time utilization records can be made the costing basis. For example, if five employees with the same job assignments actually worked for a total of 500 hours in a given month and if their collective salary was $4,000/month, the average hourly rate for their job function could be determined as follows:

$$\frac{\text{Salary/Month}}{\substack{\text{Productive Hours} \\ \text{of Work Per Month}}} = \frac{\$4,000/\text{month}}{500 \text{ hrs/month}} = \$8.00/\text{hour}$$

As we shall see in the ensuing discussions of cost centers, the determination of personnel costs can be considerably more complex than the above example. The principle of using productive time, however, will still be basic to the approach.

The problem of assigning computer operator personnel costs requires a somewhat different approach. It is normally assumed that the activities of all computer operators assigned to a computer are essential to enabling that equipment to function. This holds true even though Operator A may work exclusively at the computer console while Operators B and C attend to the peripheral needs of card readers, printers, and disk units. It is also assumed that the measure of their productivity is the extent to which the computer is or is not utilized. Hence, in this situation the measure of personnel utilization becomes equipment utilization. Assuming that a computer was productively used for 600 hours during a given month and the combined salaries of the computer operators for that month were $6,000/month, we could determine:

$$\frac{\text{Salary/Month}}{\begin{array}{l}\text{Hours of Productive}\\ \text{Computer Utilization}\end{array}} = \frac{\$6,000/\text{month}}{600 \text{ hrs/month}} = \$10/\text{hour}$$

Data Processing Supplies Costs

As data processing centers assume a greater workload and as they improve the efficiency with which they are able to use their people and equipment resources, the relative importance of supplies as a cost factor tends to increase. This is particularly true with regard to computer-form paper. To determine the cost of data processing services, the analyst must design a mechanism for recording and controlling the inventory and utilization of data processing supplies.

The primary concern of this section is: How do we record the use of supplies and apply the cost to the user of data processing services? Therefore, we must try to establish the following for any given period of time:

1. What is the quantity of supplies used?
2. What is the cost of supplies used?
3. Who used those supplies?
4. For what were the supplies used?

The problem is complicated. Some supplies such as preprinted student grade reports may have only one user, such as a college registrar's office. Other supplies such as punched cards may be used in variable quantities by various customers and in a variety of ways. Clearly, it is not feasible to handle this latter example by asking a machine operator to count the supplies used for each job. Some compromise must be made which will attempt a systematic and equitable distribution of real costs.

One way of dealing with this problem is to establish optional methods for allocating supply costs. In the case where there is a special-purpose (and hence usually expensive) supply such as the student grade report form mentioned previously, the customer may be charged for that supply either at the time that it is ordered or as it is used.

When the supply used is a standard one, the cost of supplies can be included as a function of equipment. This approach assumes that generally, as a machine such as a card punch is used more, it will use more of the supplies which can be associated with it. This means that we simply have to record total supplies used on a machine and not the amount used by each customer. There are two relatively straightforward methods for recording supplies used at the machine station. The first requires that all supplies be tagged at the time of receipt. Frequently this tagging is done with a machine-readable document. When the supplies are used up the tag is removed from the carton and placed in a small receptacle near the machine. On a regular (perhaps monthly) basis, these tags are collected and processed.

The second method is a manual one in which a supplies utilization sheet is marked. This sheet is affixed on or near a machine and has a place where both the kind and quantity of supplies used may be entered. A simple mark is made along side the appropriate supply for each unit (such as, a carton) which is consumed. The cost of these supplies will be distributed to customers in proportion to their use of the computer.

In applying the rule that the customer will be charged for supplies depending on machine utilization, we are basing our action on an assumption which is not always true. However, in applying the rule we are able to systematically and reasonably distribute all data processing supply costs to our customers. It should be remembered that any attempt to distribute all of our data processing costs is at best an estimation. What is perhaps most important is that our methodology is both reasonable and consistent in terms of costing practices.

THE DATA PROCESSING COST CENTER CONCEPT

It has become increasingly common for most large public and private institutions to establish decentralized centers of fiscal responsibility and accountability. These are called cost centers. They develop their own charge rates for services which they provide both within their organization and to outside customers. At first glance the data processing center as a whole may appear to readily lend itself to this cost center concept. The following examples deal with whether it is possible to establish rates for data processing services and whether a single data processing cost center can be set up or multiple data processing cost centers are necessary.

Example 1:
Assume that a college using data processing for academic purposes has 2500 students and spends $80,000 per year on equipment, $60,000 per year on personnel, and $15,000 per year on supplies and

expenses; then the cost of data processing could be determined as follows:

$$\text{Cost/Student} = \frac{\$80,000 + \$60,000 + \$15,000}{2500} = \$62$$

Obviously the above example does not differentiate between who uses data processing and who does not, since cost is indiscriminately distributed to all students. Nor does it account for the particular kind and extent of services which students use. It is therefore an arbitrary treatment of customers and is of little value from an accountability standpoint.

Example 2:

A college provides data processing services to both academic and administrative customers. It maintains adequate records of all its costs for equipment, personnel, and supplies. By adding all of its associated costs, it develops a per-hour cost for data processing services based upon the use of its computer. Its costing formula appears as follows:

$$\text{Cost/Hour} = \frac{\text{All data processing costs (personnel, equipment, supplies and expenses)}}{\text{Hours of actual computer utilization}}$$

This is quite different from example 1 in that it at least gives the appearance of being systematic and equitable. The approach would work very well if we could be assured that customers would use all services to the same extent that they used computer services. This is obviously not the case. For example, the student at a college or university who studies data processing may require a considerable use of the computer itself, but very little use of either systems and programming personnel, or full-time keypunch and clerical staff. In another instance, Customer A may require six months of professional programmer time for a job which is run only once a semester for one hour. Customer B may require one month of professional programmer time for a job which is run weekly for twenty minutes. If the sole criterion for payment is the amount of computer run time, Customer B will pay for a proportionately larger share of the programming cost while Customer A actually received more of those services.

One of the ways in which these and similar problems have been successfully dealt with is the practice of treating data processing services as if they represented a number of rather distinct cost centers. In most instances the three-cost-center concept works effectively. The three cost centers are:

1. Data Preparation Cost Center
2. Systems and Programming Cost Center
3. Computer Cost Center

Each of these major divisions of cost can be further subdivided and analyzed.

The Data Preparation Cost Center generally includes people, equipment, supplies, and expenses associated with the preparation of data for computer processing. Keypunching, key verifying, data recording, optical scanning, and electric accounting machine (EAM or tab) equipment processing are services normally provided by this cost center.

Systems and programming services include systems analysis and computer programming. Generally no equipment charges, save an occasional keypunch machine, are associated with this cost center. All other equipment services such as computer time for program debugging are billed to the customer directly by the cost center performing the service. In this way the need to charge-back costs internally within the data processing department is minimized.

The Computer Cost Center is generally (though it need not be) restricted to a single computer or network of computers. By attempting to individualize our computer cost center approach in this manner, we will be in a better position to evaluate specific personnel and equipment costs. Both equipment hardware and software costs (including program-product packages) are included in this cost center.

The three-cost-center approach appears to be gaining considerable support for a number of reasons. First, it enables customers to be billed specifically for services received. Second, as a management tool, it lends itself to planning, budgeting, and evaluation of data processing services by functional category.

Preparing the Billing Statement or Cost Estimate

Whether one is billing a user for actual services rendered or determining the cost for an existing or proposed system, it is necessary to present a detailed and informative data processing cost statement. In this respect, the user will want to know:

1. The particular service being referenced
    a) Data Preparation
    b) Systems and Programming
    c) Computer

2. The breakdown of people costs
   a) Rate per hour
   b) Number of hours required

3. The breakdown of equipment costs (including supplies factor)
   a) Type of equipment
   b) Rate per hour
   c) Number of hours required

4. The breakdown of special-purpose supplies and/or any billable overhead.

Obviously, this depth of information will be of valuable assistance to the user in making the right decisions about using data processing services and will assist the systems analyst in presenting comprehensive cost data as a part of any systems proposal.

CHAPTER SUMMARY

A primary ingredient in the determination of systems feasibility is to test a system for its economic impact on an organization. For this reason it is important for the analyst to consider cost data within the context of the organization itself.

The analyst examines costs which are incurred in developing, implementing, and operating a new system and benefits which result from that system. These costs and benefits are classified as (1) tangible or intangible; (2) direct or indirect; and (3) recurring or nonrecurring.

In order to determine systems costs, a number of cost categories are considered including personnel costs, equipment costs, supplies and expense costs, and overhead costs. Each of these cost categories should be considered as tangible/intangible, recurring/nonrecurring, and direct/indirect.

Cost data is frequently presented in a manner which is best described as comparative. That is, the cost of a new system is contrasted to the cost of the current system and to the cost of alternative new systems being proposed.

Two quantitative methods of cost comparison were discussed namely: (1) Break-even analysis and (2) Payback-period analysis. Break-even analysis as it is used in systems analysis is a method of determining the point at which the cost of a new system will be equal to that of the current system. The payback period on the other hand attempts to determine the point at which the investment in the new system will be totally recovered as a result of new system economies.

Data processing costs are a matter of growing importance to the determination of overall system costs. It is, therefore, important that the analyst gain considerable skill and understanding in working with these costs.

In developing and presenting data processing costs it is suggested that the analyst consider as separate cost centers data preparation, systems and programming, and computing.

WORD LIST

break-even analysis—a method of cost comparison which is used to compare the cost of new system alternatives to the cost of the current system and determine the point at which the new system will cost the same as the current system

direct cost—those systems costs which are incurred as a direct consequence of operating the subsystem in question

indirect cost—those systems costs which are incurred as a consequence of operating other subsystems which support the subsystem in question

intangible cost—a cost which is assumed to exist, but which cannot be proven

nonrecurring cost—a one-time cost frequently associated with the development and implementation of a new system

payback period—a method of cost comparison used to determine the point at which the financial returns from a new system will equal, or payback, the amount that was initially invested in the new system

recurring cost—a continuing cost frequently associated with the day-to-day operation and maintenance of the system

systems benefits—those products or improvements which result from the system and which benefit the organization

systems costs—those costs which contribute to the expense of developing, implementing, and operating a system

tangible cost—a cost for which the actual dollar amount is known or can be estimated

QUESTIONS AND PROBLEMS

1. Linda Reed, an associate systems analyst, has been asked to take charge of a new, large-scale, systems development project. To get a better sense of the project's magnitude, she assembles and reviews cost data. She learns the following:

   a. The project will cost approximately $500,000 to complete.

   b. When completed, the new system will cost about $150,000 per year to operate, compared with $125,000 per year for the current system.

   Linda believes that this information should be presented to management as part of the total systems proposal. If you were Linda's manager:

   • Would you want this information presented in the systems proposal or separately?
   • How would you like the figures to be explained?
   • Draw the kinds of graphs and charts you would like her to use.

2. Testube Pharmaceutical plans to develop a new personnel system to economize in both the personnel and payroll offices. You have been hired as a systems consultant and will meet with management next week to present your approach to making sound systems cost determinations. Prepare the topical outline and illustrative graphs and charts you will use in your presentation.

3. Larry Schucks, Director of Planning for CFO Financial Corporation, strongly disagrees with Tim Light, Project Leader for the corporation's new systems development effort in accounts payable. Larry argues that, to be useful at all, the cost/benefit information must be restricted to costs or savings in the accounts payable area. Tim believes that the project's effects on other offices must be evaluated.

   a. Do you agree or disagree with Tim? Why?
   b. What kinds of costs might Tim legitimately consider including in his cost/benefit calculations?

MINICASE—CONTEXT INTERNATIONAL, LTD.

As part of your review of the Freeflow Sludge Separation subsidiary, you discover that it is replacing an outmoded EAM (electrical and mechanical) system. You are given the following cost data on the current system and a proposed system alternative and will be asked to advise management about these systems:

Current System

Seven employees (average salary $13,000/year) are required to operate the system. Supply costs are $4,500/year. For a series of EAM and keypunch machines located in the office, equipment costs are currently $8,000/year.

Operating costs are expected to increase by an average of 10%/year in each cost category over the next 6 years.

Proposed System

The proposed system eliminates the EAM and keypunch functions, which include 3 full-time employees as well as the associated equipment. Thus, the average increase in costs over the next 6 years can be reduced to 2½%/year.

To replace the EAM and keypunch equipment, 75 hours of computer time will be purchased at $46/hour, and 220 hours of outside data entry service will be purchased at $8/hour.

One $26,000/year systems analyst will take 1 year to develop the new system. Programming costs are estimated at $20,000 for the first year and $6,000 for the second year. Training costs are estimated at $6,000 for the first year and $500 for the second year. Program testing of the total system will require 30 hours of computer time during the first year and 5 hours during the second year.

1. Prepare a comparative cost presentation in chart form for the current and proposed systems.
2. Prepare a bar graph comparing total systems costs.
3. Determine the break-even point for the proposed system.
4. Determine the payback period for the proposed system.
5. Would you favor the proposed system strictly on a cost basis? Why?

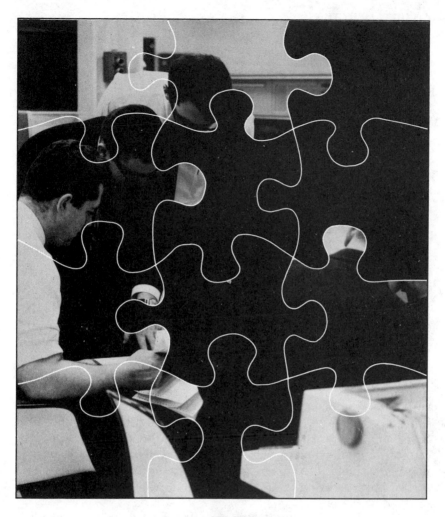

8

Systems Design—
A Structured Approach

CHAPTER OBJECTIVES

When you finish this chapter you will be able to—

- *State the advantages of the structured, top-down approach to systems design.*
- *Identify and discuss the functional design requirements that the analyst must consider in the process of systems design.*
- *Discuss the development of data dictionary concepts.*
- *Construct a data element dictionary from a series of reports and/or a narrative statement.*
- *List and discuss the three major audit considerations that must be included in a systems design.*
- *Discuss systems design requirements as they relate to online and batch input and output requirements.*
- *Discuss the role of the systems analyst as it relates to developing computer program specifications.*

Systems design is the development of the actual mechanics for a workable system. It is also the finalized master plan for problem solution. Prior to design, the attention of the analyst has been focused on how people perform their jobs, the problems they encounter, and alternative ways in which the jobs could be performed more effectively. During design, the attention of the analyst is absorbed in structuring the manner in which work will be performed in the new system.

The design of the new system must be done in great detail as it will be the basis for future computer programming and systems implementation. For example, in a new payroll system design, the analyst must know in great detail about the employee time-sheet data (which will serve as input to the system), the employee master file (which will tell us about employee pay rates and benefits), the paycheck and how it will be issued and controlled, as well as all of the procedures and guidelines that the payroll office and other personnel must follow in performing this process.

Normally, design is performed in two phases, namely, logical design and physical design. During logical design, one identifies in general terms the systems inputs, outputs, processes, and types of human interaction required by the new system. For example, the kinds of reports to be produced and the information they will contain are specified in the logical design phase. During physical design, one develops the design in greater detail and represents it as it will appear in its final form. For example, during physical design, one constructs report mock-ups for each of the previously mentioned output reports. Also, in the physical design, one indicates which computer programs and file structures are needed to support the new system. These are indicated in sufficient detail to permit program specifications to be written.

In this chapter we will first discuss structured, top-down approaches to systems design. By approaching design in a structured, top-down fashion, the systems analyst creates a framework of order within which to perform the many detailed aspects of design work.

Second, we will discuss a number of detailed design functional requirements that are of concern to the analyst in developing logical input, output, and processing specifications.

Third, a number of important areas will be discussed that are of special concern to the analyst during physical design. These include the identification and organization of data for central storage, auditable systems, output design, and operational requirements. The analyst's decisions about these matters frequently determine whether or not the system will function in the most effective way.

Also, we will make some brief comments about program specifications and development schedules, because these aspects help ensure that systems programming and implementation are timely and accurate. Because data file and data base structures are topics requiring more comprehensive coverage, they are presented separately in chapter 9.

STRUCTURED, TOP-DOWN DESIGN

The rationale for performing design in a top-down fashion is quite straightforward. That is, by starting with a global overview of the new system, the analyst, in effect, constructs a comprehensive picture of what the new system must achieve. Figure 8-1 illustrates such an overview for an order processing system. Contrast this approach with the alternative of beginning at a very low level—say, designing the customer invoice generation process. History has taught us that, when design begins at a very low level of detail (the bottom-up approach), frequently pieces are missing or the details do not fit neatly back into a wholistic system. By starting at the top, one creates a structure that ensures that sound interrelationships between systems components are built into the design at the outset.

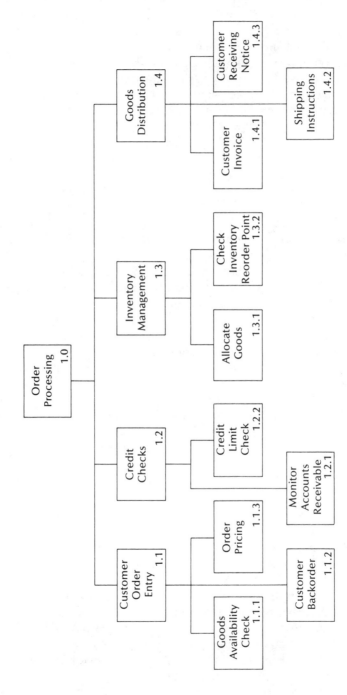

Figure 8-1 Structured overview of an order processing system

Provided that a structured, top-down approach was used during the earlier stages of analysis, the structure of the new design will have a firm base to grow on. Even though the methodologies to be employed may be vastly different from the old system (for example, online versus batch processing), it is likely that the functions to be performed will be an extension, not a wholesale replacement, of those performed by the old system. Thus, a starting point in design is to review and appropriately change the structured overview diagrams prepared during the early analysis. It is perhaps worth illustrating at this point how structured, top-down design helps the systems analyst avoid several major pitfalls that historically accompanied nonstructured approaches, namely:

- Incomplete design
- Overdesign
- Low user participation and acceptance
- Difficulties in determining project status
- Poor allocation of technical resources

It has already been mentioned that the starting point for new design should be to build on the structured diagrams developed during the early analysis. Techniques such as HIPO and, in particular, VTOC diagrams provide a quick, easy-to-understand method for reviewing what the new system should achieve. This review helps ensure that the design is complete and further establishes agreed-on boundaries, thereby making clear what the system will not do.

Because a structured overview maps out the design's total magnitude, it creates a framework for assessing the relative importance of each component to be contained in the new system. Experienced designers recognize that each additional function to be performed by a system adds to its complexity and hence to its delivery time. The systems analyst is therefore encouraged to use the structured overview as a vehicle for raising questions about the need for each function, thereby avoiding potential problems of overdesign. Naturally, all essential functions will be maintained, and the "nice-to-have" functions will be curtailed or even eliminated.

Structured methodology also creates an opportunity for increased and more controlled user participation during the early stages of design. This opportunity is further enhanced when structured methodology is used with a Team Model, such as that mentioned in chapter 1. Rather than being asked to concentrate on developing their own "laundry lists" of needs, users are asked to help construct the structured overview. As a result, they emerge from the process with a clear, nontechnical understanding of what the new system will and will not achieve. This increased participation in helping to

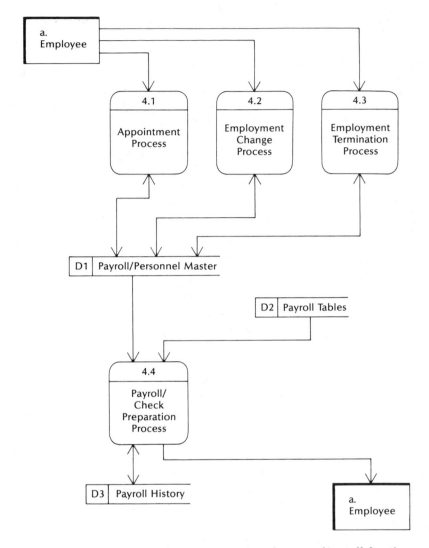

Figure 8-2   Sample data flow diagram—overview of personnel/payroll function

define the new system generally creates a higher sense of user own-
ership and therefore maximizes the opportunities for user acceptance when the system is delivered later.

Finally, note that structured design techniques help the project
manager allocate technical resources and evaluate project status.
By dissecting a complex system into its natural taxonomy of functions, the project manager can assign design responsibility to specific technical staff members for various logically contained design
components. This approach minimizes wasteful job overlap. As design
components are completed, the responsible person submits evidence

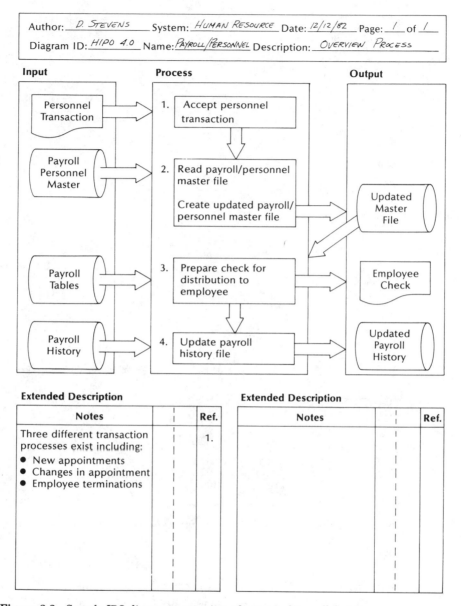

Author: _D. STEVENS_____ System: _HUMAN RESOURCE_ Date: _12/12/82_ Page: _1_ of _1_

Diagram ID: _HIPO 4.0_ Name: _PAYROLL/PERSONNEL_ Description: _OVERVIEW PROCESS_

Figure 8-3 Sample IPO diagram—overview of personnel/payroll function

of completion to the project manager in the form of a deliverable, such as a completed structure chart, diagram, and/or narrative. This deliverable gives the project manager a real sense of what has been completed, is being worked on, or has yet to be started.

Naturally, once the overview structure has been determined, it is necessary to design, in logical form, the way the new system will

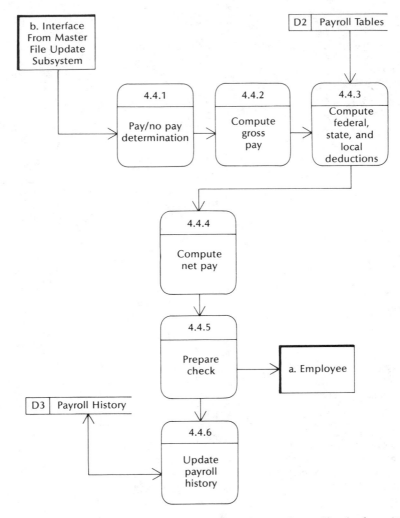

Figure 8-4 Detailed data flow diagram for personnel/payroll—check writing subsystem

perform the various functions. In the next section, we will discuss some of the considerations in the logical design of systems outputs, inputs, and processes. However, at this point it is important to note that the logical design of these important systems functions is best performed by once again employing a top-down approach. For example, as the analyst assigned to develop the payroll component of a personnel payroll system, you might have initially constructed an overview of the process by using a data flow diagram, as illustrated in figure 8-2. Alternatively, you might have used an IPO diagram, as illustrated in figure 8-3.

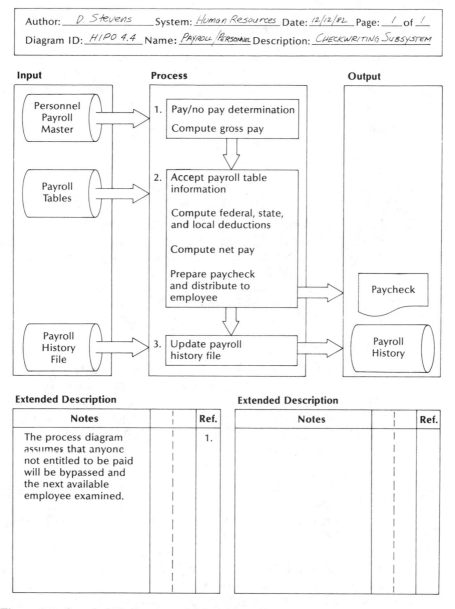

Author: _D. Stevens_ System: _Human Resources_ Date: _12/12/82_ Page: _1_ of _1_
Diagram ID: _HIPO 4.4_ Name: _PAYROLL/PERSONNEL_ Description: _CHECKWRITING SUBSYSTEM_

Input

Process

Output

Personnel
Payroll
Master

1. Pay/no pay determination
   Compute gross pay

Payroll
Tables

2. Accept payroll table
   information

   Compute federal, state,
   and local deductions

   Compute net pay

   Prepare paycheck
   and distribute to
   employee

Paycheck

Payroll
History
File

3. Update payroll
   history file

Payroll
History

Extended Description

Extended Description

Notes		Ref.
The process diagram assumes that anyone not entitled to be paid will be bypassed and the next available employee examined.		1.

Notes		Ref.

Figure 8-5 Sample IPO diagram—check writing subsystem

Regardless of which structured technique you used, in either case the approach would have been nearly the same. That is, having completed the overview data flow diagram or the IPO diagram you would have worked down to lower and lower levels of design detail, such as those illustrated in figures 8-4 and 8-5.

LOGICAL DESIGN REQUIREMENTS

It is important to learn about the alternatives and requirements for effectively designing the various new systems functions. In this section, we will introduce many of the requirements and alternatives that analysts must be aware of when defining:

• Input requirements
• Processing requirements
• Output requirements

Input Requirements

The input requirements for the new system necessitate that the analyst determine the points at which input will be introduced and the methodology to be employed in:

• Data extraction and preparation
• Data editing and error correction

Data Extraction and Preparation

Preparing input data is the process of selecting or extracting from a source document, such as an employee time sheet, the data necessary to the system. Not all the data appearing on a time sheet will enter the system. In other words, data preparation is really a screening process. Consequently, care must be taken to ensure that data which is needed by the system can be:

• Readily identified
• Easily and accurately recorded
• Readily verified
• Prepared for machine processing with little or no difficulty when the system requires data processing

Required information can be readily identified if the instructions for data extraction are clear and explicit. It can be easily and accurately recorded if a well-designed source document is available for use. It can be readily verified and processed providing that it has been prepared legibly and according to a standard format.

When the system requires that the input data be submitted for computer processing, additional steps are required to translate the source document into a machine-readable form. This generally requires both preparation and verification of the data into a machine-readable form. Figure 8-6 illustrates the procedural steps

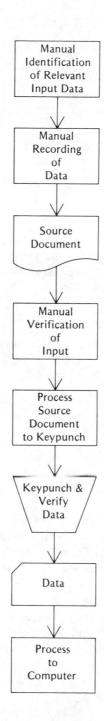

Figure 8-6   Data extraction and preparation in a computerized system

involved in data extraction and preparation when a computerized system is involved that uses conventional keypunch machines.

The preparation of data for computer processing can be accomplished in a variety of ways, including:

1. *Keypunching and verifying* Uses standard cardpunch and verify equipment.

2. *Key-to-tape data recording* Uses standard keyboard equipment which writes directly to magnetic tape and permits verification.

3. *Key-to-disk data recording* Uses standard keyboard equipment which writes directly to magnetic disk or diskette. Frequently, these systems can be internally programmed to also edit input data.

4. *Optical scanning* Uses equipment which reads, records, and verifies directly from the source document. There are many different kinds of optical scanning equipment. The requirements for preparing the machine-readable source document vary from one approach to the other.

5. *Terminal data entry* Uses any of a wide variety of computer terminal devices which permit direct entry of data into a computer system.

6. *Direct voice input to a computer* Uses highly specialized equipment which will translate the human voice into a machine-usable form.

For online systems using interactive computer terminals, data extraction and preparation can be improved by eliminating several of the tedious steps involved in transferring information from one site to another and by relying on the computer to assist in a number of ways. For example, the computer terminal may be placed at the "point of service," where the business transaction is taking place. In this case, the computer may "prompt" the terminal operator to ask the customer the minimum required questions and permit the operator to directly enter the information via the computer. This approach eliminates the need to prepare a handwritten source document by combining that step with the data entry step. Also, the computer may edit and verify the data, thus replacing time-consuming and costly manual and key verification. Figure 8-7 outlines this improved process.

During physical design, the analyst assists in the development of procedures, instructions, and forms required for effective data

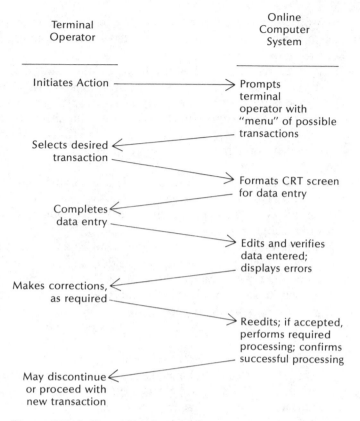

Figure 8-7 Online, point-of-service data entry

extraction and preparation. It is also likely that the analyst will be involved in training the personnel who will extract and prepare the data, and possibly in the evaluation and selection of any additional equipment, such as keypunches, which are required to support these activities.

Data Editing and Error Correction

The editing of input and the procedures required for analyzing and correcting input errors should be viewed as a single control process. It is the analyst's responsibility to design this control process in a way which is comprehensive in its detection of errors and efficient and reliable in the processing of corrections.

GIGO is an acronym which stands for *G*arbage *I*n *G*arbage *O*ut. Quite simply this means that if the system input data is in error, then the system output will also be in error. However, the truth of this does not absolve the analyst from a responsibility for the quality of the input. To the contrary, the analyst has prime responsibility for designing a system which will not permit errors to enter the system.

In addition to developing instructions, forms, and procedures which can improve the quality of input data, it is possible to specify in the systems design the need for computerized editing of all input data. Computerized editing is more extensive than that which can be done manually. Three generally accepted methods for editing input data via computer include:

1. *Missing-data test*   The input is checked to ensure that all required data is being submitted.
2. *Error-data test*   The input is checked for obvious errors such as an invalid customer number.
3. *Unreasonable-data test*   The input is compared against some predetermined standard which is considered to be the limits of reasonableness. For example, in a payroll system the number of hours worked in a week may be checked against a standard of fifty hours per week, if that is determined to be the likely maximum amount that any employee would work.

When a computer is to assist in editing, several approaches can be taken. First, the system can be designed so that specific computer programs contain the editing logic required to check for each kind of error. The payroll system in our above unreasonable-data test example used such an approach. However, sometimes the number of possibilities is extensive (for example, the product type being ordered may be one of 3,000 products the supplying firm manufactures). In these cases, it is best to design the system with special editing tables against which input can be computer-validated. This process is illustrated in figure 8-8.

Once potential or known errors have been detected, it is important to analyze these errors for their full meaning. It is very helpful to the person who is responsible for analyzing errors if, in addition to detecting the error, the computer is also programmed to provide a clear and complete explanation of the error. For example, if an input card is rejected because of data that is in error, an error report such as the one in figure 8-9 might be prepared.

In an online interactive environment, special terminal features such as "highlighting" or "blinking" can clearly identify the error for the terminal operator. These and other terminal features that assist in error detection and correction will be discussed in a later section on CRT design.

With computer assistance, persons analyzing errors can quickly determine what the problem is and initiate a speedy corrective ac-

Input Request

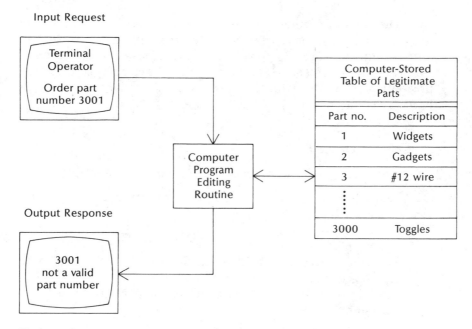

Figure 8-8   Example of the use of a computerized table for editing

Input Edit For			
Cash Up—New Cash Received			
Receipt Number	Amount	Soc Sec Number	Student Name	Error Comments
---	---	---	---	---
01446	300.00	001-32-5667	James, J.	
01447	3,200.00	002-32-5668	Valquez, J.	Exceeds reasonable amount limit
01448	760.00	003-32-5669	Bach, P.	
01449	A60.00	004-32-5670	Mulligan, M.	Invalid amount
01450	82.75	005-32-5689	Winns, S.	Invalid Soc Sec Number
01451	1,320.00	006-32-6667	Stanton, E.	
Total	2,380.00	excluding all errors		

Figure 8-9   Sample error report

tion. The amount of time required to make necessary corrections to input errors can be a crucial matter, since without all of the required input data the system's information base cannot be truly current. For this reason the analyst should design the correction procedures in a way which ensures that corrections will receive priority attention.

Processing Requirements

By the processing of data we mean one or more of the following:

1. Manipulation—sorting, combining, examining, comparing, and so on
2. Calculation—adding, subtracting, multiplying, and dividing, and so on
3. Translation—changing the data from a numeric code to an alphabetic name, or vice versa
4. Updating—adding to, deleting from, or changing the information base

The major design consideration in the processing of data is to select the proper process and to place it at the proper point in the system. A general rule of thumb is to do processing at the time when it is needed; not before. In order to illustrate the rationale for this, let us examine figure 8-10.

In reviewing this flowchart, we find that the placement of the sorting process becomes important if we consider that sorting data can be time-consuming. If the sorting were done earlier than required and if a problem occurred, then the entire system would have to be restarted and the data resorted. In the above example, it is highly probable that the only time we would have to resort the data would be if there were a problem in the sorting process itself.

Many systems require the storage of permanent information, which must be updated or kept current to reflect all changes resulting from business transactions. For example, an accounts receivable information file should reflect the most current available information regarding the outstanding monies owed the company.

In order to maintain current information the systems analyst should design methods and procedures for:

1. Adding new information to the information base
2. Deleting unnecessary information from the information base
3. Making necessary changes to the information base

Figure 8-11 illustrates some of the add, delete, and change/update considerations for an accounts receivable information base.

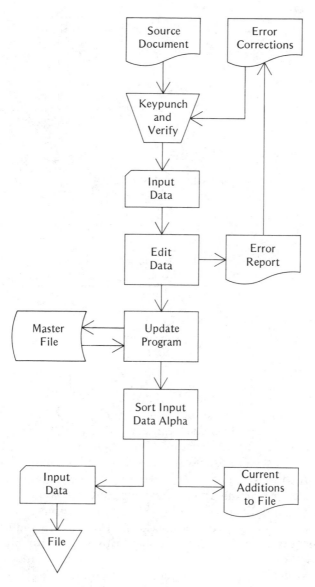

Figure 8-10   Flowchart of processing steps

There is a second kind of updating which is important to systems design. This is frequently referred to as *expanding the transaction data*. In order to speed up the data preparation process, it is common practice to store certain constant information in a computer system and to then allow the computer to update each transaction with this constant information. For example, in a computerized order processing system, the complete name and address for all custom-

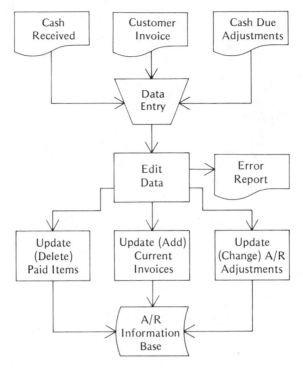

Figure 8-11   Add, delete, and change considerations in
an accounts receivable information base

ers can be stored in a computer file. At the time that an order is
received, only the customer number, which is a pre-established
unique identifier for that customer, is entered with the order transaction. As one step in processing the order, a customer look-up will
be performed and the order transaction will be updated to include
the latest known name and address for that customer. Figure 8-12
illustrates a simplified version of this updating process.

A third type of update function is the initial creation of the system information base(s). In many ways this kind of update function
is like the adding of data to an information base. Sometimes it is
possible to use the same computer programming logic to initially
create and then later on to add to the information base. However,
there are some extra design considerations involved in creating an
information base.

When a computerized information base is being created, there is a
high concern for:

1. Missing or incomplete data
2. Inaccurate data
3. Incompatible data
4. Unnecessary data

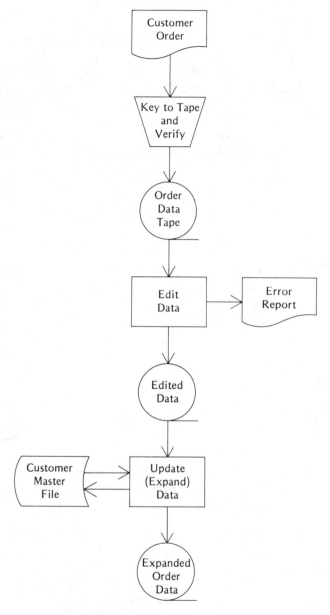

Figure 8-12 Flowchart of an updating process

The systems analyst, in designing the methods and procedures for creating a new base of information, must take these factors into consideration. At the conclusion of the create process, the information base should be both complete and accurate.

A second consideration in the creation of the new information base will be the actual timing of the create process. While this will be more fully discussed in chapter 12, which deals with systems conversion and implementation, one major point is worth noting at this time. Namely, because business is a dynamic process it is difficult to pick the right time to make a change-over to a new method for handling business information. During a change-over, care must be taken by the analyst to:

1. Limit any disruption of the normal process of business
2. Eliminate the possibility that there will be any loss or mis-handling of business transactions during the create process
3. Minimize the impact of problems which arise during the create process
4. Provide for a fallback position in case the changeover fails

Output Requirements

A system becomes useful to people through the creation of output. Therefore, the primary design requirements are that output be easy for people to use, high in quality, readily available, and retainable for possible future use.

Ease of Use

Aesthetic and content considerations are involved in output design. By aesthetic considerations we mean that an output document is:

- Attractive—in the sense of being visually pleasing
- Readable—in that the information is clear
- Organized—in that information which belongs together is placed in a common area on the output document

By content considerations we mean that an output document is:

- Brief—in that it contains only necessary information
- Comprehensive—in that it contains all necessary information
- Accurate—in that the information that it contains can be trusted as being correct
- Timely—in that the information it contains is current and suited to the needs of the business decision-making process

In our later discussions of form design and CRT screen design, some extended discussion will be given to the actual design of output formats.

Quality Control—Access to and Distribution of Output

Just as there must be designed control over the quality of input, there must also be some control over the quality of output. There are two opposing viewpoints on the responsibility for the quality control of printed output.

First, some feel that those producing the output should verify its accuracy and completeness. From this point of view it is argued that only information that has been thoroughly screened for errors should be distributed for use.

The alternative point of view is that the responsibility for the quality control function lies outside of the organization performing the work. For example, the quality control for the computer prepared accounting reports should be performed by a quality control group in the Accounting Department, as opposed to a quality control group within the Data Processing Department. Proponents of this point of view argue that such an approach:

- Is more objective (Different people are involved in evaluating and producing the work.)
- Is more informed (Close identification with the use of the output encourages a more informed assessment of its quality.)

The opposing viewpoints become resolved as one moves toward a total online environment, in which the terminal operator becomes the undisputed point for quality control.

Distribution is a second aspect of the overall concern for the quality control of printed output. In attempting to deal effectively with this matter, the analyst must obtain answers to the following questions:

1. Who should receive the output?
2. When should they receive the output?
3. How should the output be distributed?
4. Is any special handling (for example, company-confidential, registered, do-not-copy) required?

Some of the most frequently voiced user complaints have to do with either not receiving necessary output information or receiving it late. While the system may offer great advantages to people, these advantages have frequently meant that workers rely exclusively upon the system for information. When that required information is not forthcoming, they are unable to do their jobs.

One area of crucial importance in providing timely information to people is the physical delivery of output documents. Too often, careful attention is not given to creating a facility for getting output to the user on time. Regular internal mail service, postal service, and specially assigned delivery personnel are frequently employed in dealing with the problem. Because the quality of these services will vary greatly from one location to another, it is wise to specify the method which most consistently provides quick and secure delivery in a particular situation.

A similar set of problems occurs in any online environment. To deal effectively with these issues, the analyst must obtain answers to the following questions:

1. Where should terminals be located, and in what quantity?
2. How quickly should the terminals operate?
3. Who should have access to which terminals and to which data while using those terminals (terminal security)?
4. What other forms of output, if any, should be provided besides the CRT screen display so the terminal operator can work effectively (for example, "hard copy" printouts of the screens, backup reports, and so on)?
5. At what times of day must the system be accessible to the terminal user?
6. Should terminals also have stand-alone potential for word processing?

Data Retention

Data retention is an important consideration in designing an auditable system. Recognizing this, let us first examine the question of data retention as a separate design requirement. We will then build upon it in a more comprehensive way in the section on Auditable Systems.

It is necessary for the analyst to specify at what points in the process data should be stored and kept. Also, he or she should consider the storage and retention of original source documents and output documents. To this extent the analyst should be prepared to answer the following important questions:

1. What data should be retained?
2. At what point should it be retained?
3. In what form (media) should it be retained?
4. How should it be organized?
5. Who should have access to it?
6. How long should it be retained?

A number of factors may have an impact on what data is retained. In some instances there is a legal requirement that certain data be retained, as in the case of income tax preparation. At other times, as with medical case histories, it may be necessary to retain information because it is required for future use and because it is professionally ethical to do so. It is the duty of the analyst to identify what data should be retained and why. Then the system should be designed to accommodate these needs.

Data should be stored where it will be used. For example, if order data is being retained primarily to answer customer inquiries about their orders, then it should be stored at the point where these inquiries will be handled. If, on the other hand, order data is retained as a back-up in the event of system failure, then it should be retained at the point where the backup system will be initiated.

Data can be retained using a variety of storage media. Since each medium has its particular advantages and disadvantages, existing conditions will play an important role in selecting the right one for the job. Some of the alternatives and their advantages and disadvantages are indicated below:

1. Hard copy (the original or a photocopy of the original document)

   Advantages   Low cost is associated with the preparation of a suitable copy for storage. Also, in certain instances, the original document assists in meeting legal requirements.

   Disadvantages   For large quantities of data, considerable physical space is required for storage. Also, the organization and accessing of specific documents becomes a time-consuming and costly process.

2. Microfiche (a method of reducing hard copy onto 4 × 6 inch filmsheets)

   Advantages   It requires considerably less storage space than hard copy. Also, while special reader units are required, these can be obtained at a relatively low cost. The medium itself is inexpensive and less susceptible to environmental damage than many other media are.

   Disadvantages   The organization and accessing of microfiche is a potential problem area, unless some time and effort is invested. Also, information cannot be easily reorganized.

3. Microfilm (a method of reducing hard copy onto 16mm continuous film)

Microfilm

Advantages   As with microfiche, the primary benefit is reduction of bulk. However, because microfilm is not limited to a 4 × 6 inch film surface, greater quantities of related materials can be physically stored together. Also, computer output can be directly prepared on microfilm.

Disadvantages   Microfilm equipment, such as a microfilm printer attachment to a computer, is quite expensive. Also, some relatively slow-speed searching of films is required to access particular data. As with microfiche, there are difficulties if the data has to be reorganized or placed in some format other than that originally filmed.

4. Computer Storage (use of punched cards, punched paper tape, magnetic tape, and magnetic disk as the means for storing information external to the computer system)

Advantages   While there are unique advantages associated with each of these different media, these more general comments can be made: (1) some degree of reduction from very little (cards) to a great deal (disk) is possible; (2) data can be reorganized, reformatted, and retrieved at machine speeds.

Disadvantages   Machine involvement is required to locate and read data. The cost of storage and machine accessing processes is considerable.

Organization of retained data is related to the particular storage medium. In addition, however, the organization depends upon how people will want to access it. For example, if the data will be accessed on the basis of an individual's last name, then it makes sense to organize it alphabetically by last name. If, on the other hand, data will be accessed based upon its relationship to some topic (such as, architecture), then it would be wise to organize all of the data either by major topic or by cross-referencing the data to several pertinent topics. The systems analyst must specify, in the design, how retained data should be organized.

The matter of who should have access to information is emerging as a major issue of national concern. It is no longer possible for the systems analyst to deal lightly with this problem in designing a system. Rather, it is mandatory that the analyst determine some rationale for allowing specific personnel to access data and to develop methods which prevent unauthorized access. For example, in advanced computerized information systems, the accessing of information frequently requires prior entry of special "password" codes. These codes permit access to only that segment of the total information base which is needed by an individual in performing his or her job.

The length of time that data should be retained is another important design consideration. As the amount of retained information increases, so does the cost of retaining it. And yet, there are instances in which life-time retention of data becomes a fact of doing business. It is the job of the analyst to carefully analyze the rationale for both short-term and long-term retention of data with an eye toward maintaining a balance between the need to retain data and the cost.

DATA ADMINISTRATION AND DATA DICTIONARIES

During design and throughout the life of a system, it is important to maintain complete documentation about data elements and their relationships to data records, data files, reports, business transactions, and so on. For this reason, a person (or small group of people) is usually given primary responsibility for managing this information. The person is usually called a data administrator or data librarian. The repository for this documentation is called a data dictionary (also sometimes called a data element dictionary or data dictionary/directory).

The concept of the data dictionary has changed over time. Originally, data dictionaries were constructed manually and limited to the definition of data elements. At that time, each data element was explained on a single sheet of paper, as follows:

1. A brief narrative definition. For example: state-county code—a code which indicates the state or county of origin of the student.

2. A technical description. For example: a two digit alphanumeric code which is 01–60 for state students and an alpha abbreviation for out-of-state students.

3. A comprehensive list of possibilities. For example: lists that illustrate which county or state each possible code represents.

4. A list of exceptions. For example: conditions under which no county code would be recorded, such as for a foreign student.

Also, at that time the data administrator's role was narrowly defined to include these functions:

- Review data element requirements for a proposed system.
- Analyze the existing data element dictionary to ascertain whether the data currently required has been identified for other systems or has important relationships to other data.
- Develop new data element definitions and update the dictionary as required.
- Maintain current definitions for all data.
- Provide up-to-date documentation of data elements to appropriate systems, programming, maintenance, and user personnel.

Although early data dictionaries were helpful, they were costly and time-consuming to maintain, and they referred to only one aspect of systems design, namely, the data elements used. As a result, maintenance of the data dictionary did not always receive priority, so the dictionary was frequently out of date.

Several early advances in the data dictionary concept helped to make data dictionaries more central to the design process. First, data dictionaries were extended into a more comprehensive documentation support product. Second, they were automated to permit online interactive inquiry, updating, and reporting capabilities and placed in a data base environment. Thus, they became a more extensive and sophisticated tool for design as well as for maintenance of production systems. Let us now see how these more advanced data dictionaries permit relationships to be established between data element descriptions and higher-level system descriptions, thereby supporting the systems analyst's work.

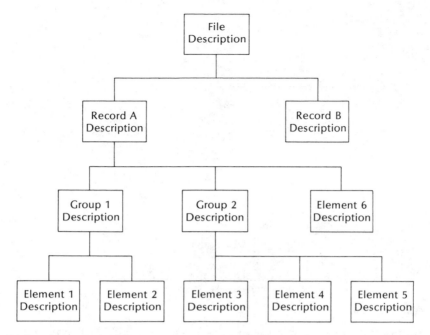

Figure 8-13   An extended data dictionary showing relationships between data element descriptions and higher-level system descriptions

The extended data dictionary is a repository for a hierarchy of data element descriptions. These can be linked to descriptions at the group, record, or file level, as illustrated in figure 8-13.

Thus, if someone wants to know what is in the data file, it is possible to obtain:

- A brief file description
- A brief description of each record type
- A more elaborate description of each record, including group-level descriptions (for example, name and address) or individual element descriptions (for example, last name, first name, and middle initial)

The extended data dictionary may store additional descriptive information. For example, it may be desirable to store system and program descriptions, as illustrated in figure 8-14.

Thus, it is possible to answer questions like, "What computer programs are used by the financial modeling system?" and "What program modules are contained in the financial forecasting program?"

The extended data dictionary is a sort of mini data base management system (DBMS). Since DBMS technology will be discussed in

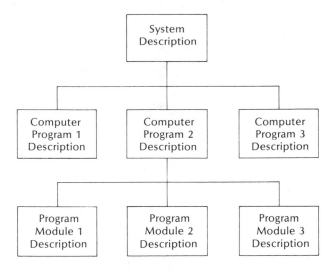

Figure 8-14   An extended data dictionary that stores system and program descriptions

the next chapter, our comments here will be brief. Suffice it to say that:

- The extended data dictionary keeps its hierarchy of descriptions in a data base (which is sometimes called a metabase).
- Access to the metabase for inquiry and updating is generally achieved online through specially prepared computer programs.
- The data dictionary has many of the other features of a DBMS, including security, recovery, and reporting capabilities.

Because the early extensions of the data dictionary concept resulted in putting data descriptions in a machine-readable form, it was to be expected that, in time, people would devise ways of accessing those descriptions as a part of writing a computer program. Indeed, today a number of software products that automatically build computer programs simply require that the "programmer" describe transactions and data requirements. Because of these and other continuing extensions, the role of the data administrator and the importance of data dictionary construction during design will grow in the future.

AUDITABLE SYSTEMS

A well-designed system should lend itself to routine auditing. It should also contain the controls necessary to ensure that, upon auditing, the system will be found to function properly. There are three major audit considerations, which will be discussed in this section:

1. Systems Controls
2. Audit Trails
3. Systems Recovery

Systems Controls

Auditors will want to know the location of systems error-control points and the reasons for selection of these points. Therefore, it is important during the design of a system to evaluate the selection of control points on the basis of:

1. Frequency of error
2. Magnitude of the error
3. Cost of error detection
4. Timing of error detection

Items 1 and 2 above are quite straightforward in their meaning. That is, as the potential for error increases or as the potential consequences become more drastic, then there is a greater need for control. By identifying points where the potential for errors and the consequences of errors are great, the analyst is frequently able to design effective error-control procedures at the proper point in the system.

There is a cost associated with effective error control. The cost of error control should be kept well below the cost which could be anticipated if errors were not detected. In other words, the analyst attempts to develop error controls which provide the required level of protection at a reasonable cost.

One particularly common approach which limits the cost of error-control procedures is called sample testing of the system. Rather than control each and every transaction, only a relatively few transactions are carefully screened—say, one in a hundred.

Another common approach is to provide some mechanism which will perform a more general check of the system. A manual batch total may be kept which accumulates all of the transactions (such as cash collected) for a given period of time. This batch total can be compared later to a second accumulated total prepared by the system. A discrepancy between the batch total and the system total indicates that an error has occurred in one or the other. Similar kinds of gross checks of the system can be performed by counting either transactions or source documents, both prior to processing and as a function of processing.

The fourth design factor associated with systems controls is the timing of error detection. In other words, the analyst must determine where to control for errors based upon the importance of detecting a particular error early in the process. Not all errors need to be detected immediately, and it may be easier and less costly to detect them later on. For example, in computerized systems extensive error checking is probably best accomplished by the computer itself, rather than by the personnel receiving or preparing the original source document.

Audit Trails

An *audit trail* is a series of designed provisions which permit a person to follow and verify the systems process. It is desirable to be able to either start at the beginning of the process with a transaction(s) and follow it through to completion, or to take results and backtrack through the process to the beginning source documents.

The following illustrates some of the materials which are required in constructing an audit trail:

    I. Source Documents
        A. Retained as evidence of the transaction
   II. Manual Batch-Total Slips
        A. A summary total for all transactions included in a batch
  III. Computerized Input Summaries
        A. A computerized summary total for all transactions included in a batch. This includes a computer-prepared batch total which can be compared to that which was manually computed.
  IV. Update Transaction Listing
        A. A computerized listing of all updates to computer files
   V. Computer-Prepared Error Listings
        A. Listing of all input which was not permitted to enter the system
        B. Listing of all transactions which did not successfully update the computer files
  VI. File Reconstruction Procedures
        A. Back-up storage of files and transactions which enable files to be recreated as required
 VII. Printout of Files and Records
        A. Ability to print out the complete file
        B. Ability to selectively print out a particular record
VIII. Final Reports
        A. Retained copies of final reports produced by the system

In addition to the advantages of audit trails for performing a valid audit of the system, they also have some important internal benefits. For example, they allow personnel to detect discrepancies and to trace back to the source of the error. They also enable personnel to respond to routine requests for information regarding a specific transaction. (For example: What happened to Customer Y's order?) Lastly, in the event of a system failure, they frequently provide sufficient material for speedy and accurate recovery.

Systems Recovery

In the previous section we indicated that a well-constructed audit trail can ease the problem of systems recovery. Systems recovery means much more than simply having a good audit trail. It is in itself a major element in systems design and is concerned with designing into a system the following:

1. Systems restart capability
2. Systems processing alternatives
3. Systems security

Systems recovery is concerned with those actions which should be taken in the event of systems failure. This is perhaps most evident in item 1 above. That is, in the instance of systems failure there must be explicit directions and procedures outlining the steps to be taken to restart the system.

Assume that the Meridian National Trust Company would like to maintain a large-scale computer in its central banking office. Furthermore, assume that there are ninety-four branch offices of the bank, and these offices will operate 300 computer terminals and be linked directly to the central office computer system.

The systems personnel assigned to this problem could clearly not ignore the possibility that the system, once installed, might fail. In order to deal with this very real possibility, it will be necessary for the analysts to determine:

- The kinds of failure which might occur (hardware, software, program)
- The potential consequences of the failure (lost data, adulterated data, and so on)
- The specific actions required to restart the system (reload data files, reenter previous transactions, and so on)

At the Meridian National Trust Company, there may be times when the system failure is severe and/or the systems restart procedures themselves are inadequate to deal with the problem. Because of this possibility the analyst must consider alternatives. Clearly the banking business cannot stop simply because of a primary system failure.

To deal with this problem it is common to design one or more back-up systems which can be placed into operation in the event of primary system failure. There is a wide range of back-up alternatives varying in both complexity and cost. For example, Meridian National could have used a completely manual system wherein all banking transactions were recorded manually by bank tellers until the system problems could be corrected. When the system was operational again the transactions could be entered into it as if they had just occurred.

The other extreme in back-up systems would be a second or even a third identical computer system waiting to go into action in the event of primary system failure. Between these two extremes there exists a broad spectrum of possible back-up alternatives. A comprehensive systems design will consider both restart and alternate processing capabilities.

Systems security is also mentioned at this time since this is one of several points where it is of obvious importance. The ability to recover from systems failure and in some instances to prevent it is in large part a function of the care which goes into the design of systems security. Some of the concerns which should be included in the initial design are:

- Provisions for the physical security of equipment and input/output documents
- Control over access to the system and system information
- System checks which prohibit or discourage abuses by authorized personnel
- Adequate methods for the storage and retrieval of historical and back-up information files (See chapter 12 for a more detailed treatment of this topic.)

FORMS REQUIREMENTS

In business, one of the most important means of communication is the business form. It is common to find, even in the smallest business, that many different forms are used. This popularity stems from the fact that a well-designed form is:

1. *Informative*   It contains information which is required for making a particular business decision.

2. *Specific*   It contains *only* that information which is required for making a particular business decision.
3. *Easy to complete*   Very little training is required to enable a person to complete the business form.
4. *Easy to use*   The completed form readily lends itself to a particular business use.

Some of the qualities which make business forms popular can also create difficulties for a business. In particular, the fact that business forms are specific means that the number and variety of forms can, if uncontrolled, become excessive.

Forms Control

Designing effective controls for the use of business forms is an important aspect of the analyst's work. Indeed, the analysis of existing forms may be a primary systems assignment. Whether it is the analysis of existing forms or the development of new ones for a new system, there are a number of forms control considerations required of the analyst.

1. Form identification—All forms should be clearly identified in a way which describes their intended use. The use of both title and form number is generally employed.

2. Date of issuance—All forms should carry with them an indication of when they were issued. This makes it possible to identify the latest version or revision for a given form.

3. Approval of new/revised forms—A procedure should be developed for the approval of all new or revised forms. This will control the unauthorized proliferation of business forms.

4. Identification of person and/or office responsible for preparing the form—Requiring an identifier of this type on the form helps to ensure that the form will be completely and legibly prepared.

5. Disposal of forms—Procedures for the proper disposal of forms after they have fulfilled their purpose is particularly important when confidential or security information is involved. The use of paper shredders, incinerators, and other disposal methods should be planned when appropriate.

The above stated control considerations apply to the use of all business forms. In addition to these control considerations, there are a number of specific design requirements for input and output forms.

Input Form Requirements

The input form is the primary means whereby data is entered into a system. Hence, its design will have an impact upon the accuracy and reliability of input data. The form must:

- Lend itself to being filled out quickly and accurately
- Lend itself to further processing

Whenever possible, the form should contain the necessary instructions for correct completion. For example, if a person's name is required, last name first, then the instruction LAST NAME FIRST should appear in an appropriate place on the form. Similar instructions such as LIMIT TO 24 CHARACTERS and START WITH MOST RECENT EMPLOYMENT can be easily built into the design of a form. When an input form requires lengthy or involved instructions, they should be placed together either at the top or on the reverse side of the form. Separate instruction sheets should be avoided whenever possible.

Sometimes it is wise to use a check-list approach to forms design. For example

> *Instead of asking:*   Sex_____
>
> *Ask:*   Sex: Male   □
>   Female □

By using the check-list approach people are able to complete the form quickly and with a minimum of errors due to misunderstanding and misspelling.

In a computerized system the primary use of an input form is as a source document. Therefore, it is essential that the input form be designed in a way that lends itself to data entry requirements. By placing information on the input form in the same sequence required for data entry, the whole task of entering and verifying information is simplified, and the potential for errors is greatly reduced. Figure 8-15 illustrates proper and improper design of an input document for data entry. Generally speaking, from a well-designed input form, the terminal operator can enter the data while reading the form from top to bottom and left to right, just as one reads a book or magazine.

Output Form Requirements

Earlier in this chapter two aspects of output form design were discussed, namely:

- Aesthetic considerations
- Content considerations

At this point it is necessary only to add to those previous comments the importance of also considering the intended use of the

IMPROPER DOCUMENT LAYOUT

(3) Name _____   (1) Soc Sec Number _____

(5) Home Address:

        Street and Number _____

        City _____

        State _____

        Zip _____

(2) Birth Date _____/_____/_____   (4) Curriculum _____

PROPER DOCUMENT LAYOUT

(1) Soc Sec Number _____   (2) Birth Date _____/_____/_____

(3) Name _____   (4) Curriculum _____

(5) Home Address:

        Street and Number _____

        City _____

        State _____

        Zip _____

Figure 8-15   How-to and how-not-to design an input document for data entry

output documents in its design. All information appearing on the output form must be fully explained. This can be accomplished using the following:

1. Descriptive heading information
2. Explanatory comments
3. Underlining and the use of special characters to draw attention to significant data

To the user of a system, it is the output (and the form of that output) which is the basis for judging the system. If the output is easy to use, then user opinion of the system will remain positive.

FORMS DESIGN

The analyst must design forms in a way which maximizes their effective use. The actual specifics of forms design are usually performed with the aid of a forms-design consultant—particularly when the form will be used on a volume basis. Most business-forms supply companies will provide these consultant services at no extra charge, providing that the analyst has developed a rough sketch of the design layout. The remainder of this chapter will concentrate on some of the alternative types of forms available from most suppliers.

Preprinted Forms

It is possible to obtain forms preprinted with heading and explanatory information. Some of the advantages of these preprinted forms include:

1. Professional printing is more attractive and more readable than computer print.
2. A wide variety of print types are available and they conserve space and permit accenting of headings as required.
3. Company or department seals and insignia can be placed on the form, thus identifying it and making it more attractive.
4. Multiple copies of the preprinted form are generally color-coded, thus standardizing the form's distribution.

There are two major disadvantages to the use of preprinted forms. First, they are generally far more expensive than stock (nonprinted) forms. Second, required changes to the form make it obsolete. Generally speaking, the analyst will use preprinted forms only when the information to be printed on the form is not subject to frequent change and when there is a premium attached to having an exceptionally attractive document, such as when it will be sent to a customer or to top management.

Multiple Forms

Business forms are usually required in more than one copy. Some of the methods for dealing with this requirement include:

1. *Copying machines*   Many of these machines now permit the reduction of the document and back-to-back (reverse side) copying.

2. *Carbon paper*   It is available either as one-time or reusable.

3. *Chemically coated paper*   Sometimes called "carbonless paper," it allows copies to be made as printing pressure is applied to the uppermost copy.

4. *Carbon-backed paper*   A carbon back is placed on part or all of the reverse side of each sheet. This can be arranged to have different portions of the original form appear on each of the copies. A more complete description is given in the discussion of Instant Mailer Forms.

5. *Two-up printing*   This is frequently used in preparing computer output. Two copies of the same form are attached. The computer prints both forms at the same time.

6. *Routing slips*   As an alternative to individual copies, a single copy is routed to all parties who should see it.

Instant Mailer Forms

Computers are fast and efficient in the preparation of printed output, such as customer invoices and monthly billing statements. Frequently the distribution of this output becomes a bottleneck when stuffing and addressing envelopes is involved. With carbon-backed paper, the machine can print information in a sealed envelope with only the name and address appearing on the outside. While this type of form is expensive, the added cost is offset by the elimination of envelope stuffing and addressing cost. Also, there is the added advantage of minimal delay in distributing the output.

Turnaround Documents

At times, it is possible to use the same form as both an input and an output document, in which case it is referred to as a *turnaround document*. The concept is quite straightforward. It includes the following considerations:

1. It must inform the user of what is required—for example, the amount of money due.
2. It must contain required information to process it as a series of input transactions—for example, the customer account number and the appropriate invoice numbers.
3. It must provide an area for the user to make necessary entries—such as the amount being paid.
4. It must be designed so that it is useful as output and also lends itself to being used as a source document.

Figure 8-16 is an example of a turnaround document which might be used in a student accounts receivable system for a college or university. Note in this example that the turnaround document also contains a history of all previous partial payments. Consequently, the form can, if it is produced in triplicate, serve as a student receipt, a temporary accounting office record, and an input form.

CRT SCREEN DESIGN

The design of effective input and output screen displays for business systems is substantially different from the interactive computing mode that most data processing or computer science students are routinely exposed to. There are many sound reasons for this difference. First, the businessperson interacting with the computer has little or no knowledge of the technology that supports the input and output functions. (In many instances, the terminal operator is a retrained typist or clerical person.) Second, the transaction environment is such that the same person may process a variety of transactions (for example, placing an order, canceling an order, checking the status of an order, and so on). If the transactions are hard to execute, it will be hard for the terminal operator to remember all the different instructions that may apply. Finally, terminals are placed in areas of high transactional activity, and their successful operation is essential to completing business transactions. Therefore, the speed of execution can correlate directly with the volume

First Semester	Power of Attorney	Resident	Meal Plan	Resident Assistant	EOP	Cont. Ed.	Vocational	Action & Date	Account Closed
	*	*	7						*

Local Address: **Porter**
642-2
6632

Account Balance: $0.00
Total Credit: $0.00
Balance Due: $0.00

PAYMENT PROFILE

Item	Billed	Paid	Credit	Balance Due
Tuition	325.00	325.00	0.00	0.00
Coll. Fee	12.50	12.50	0.00	0.00
Grad. Fee	0.00	0.00	0.00	0.00
Room	325.00	325.00	0.00	0.00
Board	305.00	305.00	0.00	0.00
Activity	35.00	35.00	0.00	0.00
Alumni	2.50	2.50	0.00	0.00
CTL	0.00	0.00	0.00	0.00
Insurance	7.75	7.75	0.00	0.00
TOTALS	1012.75	1012.75	0.00	0.00

Cash Overpayment	
Excessive Credit	

● THIS FORM IS YOUR RECEIPT

RECEIVABLES

Code	Receipt No.	Amount
B	02166	687.75
C	09565	225.00
D	09799	100.00

EXPLANATION OF CODES

A	Deposit	H	Veterans Benefits
B	Cash	J	Other Scholarship
C	SIP	K	EOP
D	SUS	L	Other
E	EOG	X	Local Refund
F	NDSL	Y	State Refund
G	Re-Hab	Z	FSA Refund

● Payment by check subject to clearance by bank.
● Bring this form with you when making inquiries regarding your account.

PAYMENT UPDATE

	Type	Amount	Receipt
B	CASH		
C	SIP		
D	SUS		
E	EOG		
F	NDSL		
X	Local Refund		
Y	State Refund		
Z	FSA Refund		

Grad. **Fee** Paid

Cashier	Date
	/ /

Figure 8-16 Turnaround document

of transactions that can be performed at each terminal. The potential benefits are measurable in such terms as fewer people using fewer terminals to serve more satisfied customers.

Let us examine how the systems analyst can design a CRT environment that facilitates:

- Ease of use for the nonsophisticated computer user
- Varied transactional processing with minimal memorization of instructions
- Higher transactional processing volume per terminal operator

Ease of Use

One of the most common approaches in designing easy-to-use CRT screen displays is the fill-in-the-blank approach. The analyst simply formats the initial input display so that all the required data elements are clearly labeled and a space is provided for data entry, as illustrated in figure 8-17. Some types of computer terminals enhance this process by permitting the data labels to appear in inverse video (black letters on a white background) while the data themselves appear in regular white letters on a black background. For further ease in moving from one input field to another, the terminal is equipped with a "tab" key similar to that on a conventional typewriter. This feature is particularly helpful in correcting or updating displayed data. For example, the operator can simply "tab" over to the data element that needs to be changed.

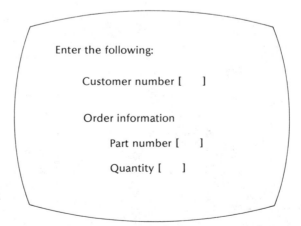

Figure 8-17 Fill-in-the-blank approach to CRT screen design

There are other ease-of-use design aspects similar to those mentioned in our earlier discussions of input requirements. That is, the display should be aesthetically designed, and all descriptions and error messages should be meaningful, clear statements. In other words, care should be taken to avoid symbols and overabbreviation. Normally, the initial input display is shown again, with the error message at the bottom of the screen. Figure 8-18 illustrates the proper formatting of a terminal display in which an error has been detected.

Some optional physical characteristics of terminals also enhance ease of use. These include:

- Display intensity adjustments that reduce operator eyestrain
- "Blinking" capability, which can highlight errors
- Tiltable screen and adjustable chairs, which give the operator more physical comfort

Minimal Instruction Memorization

There are basically two approaches for minimizing the amount of memorization required to operate a system. Each has advantages and disadvantages, depending on the specific work situation. The analyst determines which approach is better suited to a particular situation.

First, the operator instructions are commonly stored on a computer file, and a single instruction (frequently called a HELP

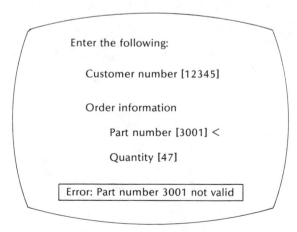

Enter the following:

Customer number [12345]

Order information

Part number [3001] <

Quantity [47]

Error: Part number 3001 not valid

Figure 8-18   Display of data entry errors

instruction) retrieves the appropriate stored instructions when they are needed. The HELP instruction approach is particularly useful when transactions are used only infrequently. However, this approach suffers because looking up instructions and reading how to process a transaction is time-consuming. Certainly we would not want the terminal operator to spend large portions of the day looking up instructions on how to perform various transactions. For this reason, when the operator must handle a great variety of transactions, a "menu" approach is generally used.

The menu approach is illustrated in figure 8-19. Basically, the approach is as follows:

1. A menu of possible transactions is displayed, which the operator selects from.
2. After the selected transaction is processed, the computer system formats and displays the appropriate input transaction.
3. The operator simply fills in the required data and retransmits the screen.

In practice, a procedure may be complex enough to require several levels of menu selection by the operator. For example, the first menu may offer order, cancel, and status inquiry transactions; for an order transaction, the second menu may offer cash order, charge, and credit card transactions.

Although the menu approach has many advantages, it should be employed judiciously, because an overly complex and rigid menu process takes time and can be frustrating to the terminal operator. In other words, if the operator will probably learn some of the

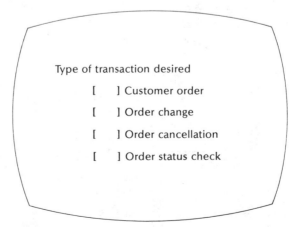

Figure 8-19   Menu of customer order transactions

instructions over time, the systems design should permit the operator to short-circuit the menu process and enter the transaction directly.

Improved Processing Speed

The major factors in processing transactions quickly from the terminal are processing or computer time, data transmission time, and data entry time. In this section, we will address only the data entry and data transmission aspects.

Some of the ways to reduce data entry requirements include:

1. Designing the screen display so responses can be abbreviated (for example, entering "Y" instead of "Yes")

2. Designing the screen format so the order of data entry is consistent with the business transaction; this feature eliminates unnecessary "tabbing" around the screen

3. Designing the screen format so data that can be changed are "unprotected" and data that cannot or should not be changed are protected

4. Using program function keys, which are available on many of the more sophisticated terminals and can literally trigger a transaction when a single button is pushed

5. Using terminals that have additional application-specific features (for example, a number pad can be helpful for accounting systems)

Quite apart from the amount of time the computer uses to process the business transaction, a noticeable amount of the total transaction time is expended on transmitting data between the terminal and computer. In a business transaction environment, we are frequently transmitting whole screens filled with data across the communications lines. For this reason, it is usually important to employ reasonably rapid data communications (between 4,800 and 9,600 baud) in advanced business systems. To do otherwise produces longer waits at the point of business.

PROGRAM SPECIFICATIONS

When designing a computer based system, the analyst must identify those areas requiring computer programming and provide a list of specifications which detail the problem for the programmer. The detail with which program specifications are developed varies greatly from one organization to another. In particular, some organizations require that an analyst prepare detailed program flow-charts of the problem and that the programmer simply code from those charts.

It has been a basic contention of this text that the analyst has numerous important systems functions to perform. As a corollary viewpoint, we add that the computer programmer is also a professional. Indeed, it is assumed that the programmer is the one most skilled in making decisions about how to write computer programs most effectively. Consequently, the analyst should work out some optimal level for specifying to the programmer what should be done without also performing major aspects of the programming function.

One of the often-cited advantages of structured, top-down approaches to design is that they ease the burden of developing good program specifications. Recall that, in using a technique such as HIPO, we first dissected the system into its logical components or functions. Then we identified the inputs, processes, and outputs for each of those functions. Consequently, our HIPO package provides the first important component in our program specifications. We may wish to augment this component with a narrative description for each function.

Detailed descriptions of the various data elements, data records, and data files are also provided as a part of the specifications. Similarly, sample reports and the sequence of their preparation will be important.

For particularly crucial logic problems or when definite parameters exist, detailed flowcharts, decision tables, and narratives may be required. For example, if one of the functions of a student record system was to produce a "Dean's List Report," it might be wise to include in the program specifications a criterion for dean's list status, such as:

A student will be determined to be on the Dean's List if the following criteria are met:

1. Student attends full-time. That is, the student is currently enrolled in at least twelve hours of study.
2. Student has for this semester achieved at least honors status. That is, the student's semester quality point index is at least 3.0 on a 4.0 scale.
3. Student has not failed or failed to complete any course in the current semester. That is, the student does not have any F, U, or I grades this semester.

When specific program functions are not obvious, they should be listed. For example, the analyst may wish to specify how data should be edited, how errors should be reported, how data should be accessed, and how it should be stored. However, in many of these instances, unless the analyst is also an expert programmer, he or she may be wise to hold some preliminary discussions with someone who is an expert.

By and large, the purpose of developing program specifications is to minimize the need for consultation between analyst and programmer during the programming phase and also to minimize the possibility that inappropriate or inaccurate computer programs will be developed.

DEVELOPMENT COMPLETION SCHEDULE

Once the comprehensive design of a system has been detailed the analyst is in a good position to estimate various project completion dates with considerable accuracy. Chapter 11 on Project Management and Project Control goes into this topic. Developing of a schedule for project completion is an integral part of systems design. Not all systems analysts will have responsibility for this aspect of the design. It is usually the project leader of the design team who develops the completion schedule.

It is customary to include systems cost data as a portion of the development completion schedule. The importance of initially developing this cost data as a part of the feasibility study was previously mentioned. During the design phase the analyst redevelops that cost data with greater accuracy. Knowing in detail what will be required to develop and implement the system will be the key to developing a comprehensive cost picture, such as that outlined in chapter 7.

STRUCTURED WALKTHROUGHS

The structured walkthrough process began with the advent of team approaches to structured programming. At that time, the process emphasized the importance of changing computer programming from a private, individual art form into a public practice.

Normally, the programmer first prepares a program design document. This document could take many forms, but the most widely used form involves pseudocode—an English-like simulation of actual programming statements.

The pseudocoded program is then distributed to a review team, which studies and develops comments about it (generally, this is done a full day before a review team meeting).

The next step in the process is to hold the structured walkthrough meeting. At that meeting, the programmer walks the entire team through the program design. The team comments on the design, pointing out "bugs," or potential problems. Also, the team comments on the quality of the program's structure (for example, whether or not it uses structured concepts, is easy to follow, is readable, and so on). In the walkthrough session, the team members do not advise the programmer on "better" or more efficient program routines to use. Rather, the session is used as an opportunity to ensure that a well-conceived, structured program with a minimum of bugs will be implemented.

Whenever changes are required as a result of the walkthrough, the process is repeated after the changes have been made. This repetition is a most important (and sometimes overlooked) aspect of effective walkthroughs, because new bugs might have been introduced as a result of the changes.

The structured walkthrough is generally used once again after the actual program coding process has been completed. In such instances, the structured walkthrough is generally called a structured code review (to differentiate it from the review of program design).

The consequences of using the structured walkthrough are considered highly beneficial, in that they lead to:

1. Well-structured, easily understood programs
2. Virtually bug-free programs that frequently run correctly after one or a few tries
3. Better-documented programs that adhere to sound programming standards
4. Programs that are more readily maintained by people other than the original designer

The most frequently vocalized complaint about the structured walkthrough is the amount of time required to execute it. Although the walkthrough adds time to the front-end process of developing a computer program, programming managers generally agree that this investment of time at the front end is but a small fraction of the time traditionally spent in trying to maintain and modify poorly designed and poorly implemented computer programs.

The concept of the structured walkthrough as it applies to computer programming is of major interest to the analyst, who will have to interact with programmers and may, in fact, be a member of the walkthrough review team. However, the concept of the structured walkthrough also has relevance for systems design, as well as for program design.

That is, the analyst should plan for periodic review of various deliverables associated with the design. Necessarily, the team that reviews the design components would be a composite of user, management, and technical personnel. Several of the practices associated with program walkthroughs seem particularly suited to systems design reviews. First, the practice of sketching out the design and distributing it to a review team before a meeting should be maintained.

Second, the review team should focus on the accuracy and completeness of design products, instead of the methodology used to produce them. (Too frequently, we find people other than the designer becoming involved in decisions about how information is stored, accessed, and so on.) The review team's primary function is not to dictate systems design, but rather to review the design for obvious omissions, bugs, and so on.

Finally, the procedure of repeating the walkthrough if and when changes are required should be followed to ensure that the changes do not result in new problems.

On balance, using the walkthrough methodology throughout the design process should help create a structured environment in which user, management, and technical data processing personnel can help the systems analyst achieve systems design products of high quality.

CHAPTER SUMMARY

Systems design is the process of developing the actual mechanics for a workable system. As such it represents in final form a detailed master plan for problem solution.

The first phase of design is usually called logical design. Because logical design includes the more general, nontechnical description of desired systems inputs, processes, and outputs, a high degree of user involvement is both possible and desirable during this phase. Physical design is the second phase of the work. During this phase, the analyst translates systems requirements into a form that is ready for computer programming.

The preferred approach to both logical and physical design uses structured, top-down methodology. By using this approach, the analyst avoids many of the pitfalls historically associated with nonstructured and bottom-up approaches.

Once the overall system structure has been established, the analyst gives some in-depth attention to the design of system inputs, processes, and outputs.

During design and throughout the life of a system, it is important to maintain complete and current documentation about data elements and their relationships to data records, data files, reports, and so on. Recent and continuing advances in data dictionary technology have made the data dictionary increasingly central to the design process.

A well-designed system should lend itself to routine auditing. As a part of this function, the analyst builds systems controls, audit trails, and system recovery capabilities into the system design.

The business form is one of the most important means of communications within a business organization—hence, within business systems. A system should be designed to maximize effectiveness in the use of forms and to efficiently control their use. Designing the right form for the right job is an important aspect of systems design since these forms represent the primary points of contact between users and the system.

The growing prevalence of online systems has created a need for the systems analyst to concentrate on the many unique aspects of CRT screen design. In particular, the analyst must be concerned with designing terminal-based systems that:

- Are easy for nonsophisticated personnel to use
- Assist the terminal operator by minimizing the number of instructions to be memorized or looked up
- Permit a high volume of transactional processing per terminal operator

As a part of systems design the analyst must determine those points in the process which will require computer programming. The analyst will have to express these programming requirements in the form of a set of detailed instructions or program specifications. While the level of detail required in these specifications varies from one organization to another, it is suggested that they minimize:

1. The need for constant consultation between analyst and programmer
2. The possibility that inappropriate or inaccurate computer programs will be developed
3. The possibility that the analyst will be unnecessarily misdirected away from systems to programming responsibilities

Refinement of project completion and cost estimates is possible as a consequence of the design process. It is important that the project leader determine these refined estimates during systems design.

The structured walkthrough methodology, which began during the early days of team approaches to structured programming, helps the systems analyst coordinate the design and coding of computer programs. This methodology can also be used during systems design to assure consistently high design quality.

WORD LIST

audit trail—a series of designed provisions or capabilities which permit a person to follow and verify the systems process

data administrator—the person, sometimes called the data librarian, responsible for maintaining complete, up-to-date documentation about data elements and their relationships to data records, data files, reports, and business transactions

data dictionary—a facility, sometimes called a data element dictionary or data dictionary directory, that supports the maintenance of current documentation about data, programs, and other aspects of large, complex systems

data element—the smallest component of meaningful data which can be represented

data extraction—the process of selecting from the environment that input data which is necessary to the system

GIGO—an acronym which stands for *Garbage In Garbage Out*

logical design—a general, business-oriented description of a system and its input, process, and output requirements

microfiche—a four-by-six-inch filmsheet copy of printed materials

microfilm—a 16mm continuous film copy of printed materials

physical design—a detailed, technical data processing description of hardware, software, and data base (or file) organization requirements

reasonableness test—the examination of data to determine that it is within a prescribed or predetermined set of acceptable limits

structured walkthrough—a method in which a review team works with the computer programmer or systems designer to critique products before their implementation

systems design—the process of developing a detailed master plan for dealing with a systems problem(s)

transaction menu—a list of alternative transactions that is displayed on the CRT screen and used to prompt the terminal operator

turnaround document—a business form which is used as both an input and an output document

QUESTIONS AND PROBLEMS

1. Data Element Dictionary Problem:

   Elmo Jackson is a systems analyst at the Applied Educational Institute of North America. During a study of the Student Records System for that college, Jackson developed the following descriptions of office information needs.

   College Registrar: The office requires information regarding many aspects of the student's academic life, including courses previously taken or currently enrolled in (course and section number and course name), credit hours, grades received, grade point average, and academic status (probation, honors, high honors, and suspended).

   It should also be noted that the Registrar's Office is the official point at which students may withdraw or report name and/or address changes.

   Admissions Office: The office requires information regarding the student's high school average, College Entrance Examination Board test scores, state/county of origin, and college program (curriculum) preference.

   The college must be able to correspond with the student, the student's parents, and the high school guidance office.

   Based upon an evaluation of the student's high school and standard test performance, the admissions office will either accept, conditionally accept, or reject the student application.

   College Bursar: The office requires a variety of information in order to properly determine how to bill a student and what if any refunds should be paid to a student.

   All freshmen students are required to pay an application fee of $50 and a freshmen orientation fee of $15. The amount of tuition which a student is required to pay depends on the following:

	In-State	Out-of-State
Upper division (completed 57 or more credit hours)	$700/sem.	$1,050/sem.
Lower division (completed less than 57 credit hours)	$450/sem.	$ 800/sem.

Had you been the analyst, how would your approach have differed from that used by Jackson?

Using Jackson's descriptions, develop a tentative list of data elements which you think should be included in a data dictionary. For six of these tentative data elements, describe the approach you would like to take in defining them and the kinds.

2. Design a preformatted CRT screen (or screens) for first-time capture of basic employee data, including:
    * Social Security number
    * Employee name
    * Employee home address
    * Employee office building and room number
    * Home telephone number
    * Work telephone number
    * Date of birth
    * Sex

    You can select any reasonable data element size and the order in which elements should be entered.

3. Design menu screens for the following:
    1.0 Orders
        1.1 New order
        1.2 Order change
        1.3 Order cancellation
        1.4 Order inquiry
    2.0 Billings
        2.1 New billing
        2.2 Revised billing
        2.3 Billing inquiry
        2.4 Dunning notice
    3.0 Collections
        3.1 Cash or check
        3.2 Credit
        3.3 Bad-check process
        3.4 Status inquiry

4. As a member of the Data Administration Unit at Freehold Industries, you have been asked to describe a data dictionary product that permits the establishing of data element relationships to data records and to computer programs. The data dictionary also permits one to establish data records relationships to data files and ultimately to a data base. Computer programs will be linked to subsystems and to systems. Draw a diagram illustrating these data dictionary relationships.

5. As the lead systems analyst on a new personnel system project, you wish to monitor the process strategy. How will you monitor the process to ensure that the work is being performed according to your specifications?

6. Consider a bill that you (or someone in your family) received recently. If it is a turnaround document, what does it do well? Redesign it to overcome inherent problems. If it is not a turn-around document, redesign it as a turnaround document.

7. Design a payroll system that pays hourly and salaried employees biweekly, with quarterly reporting to the federal government on FICA deductions and contributions. A payroll register will be produced when the paychecks are generated. In your design, be sure to illustrate how the following functional design requirements will be met:

    a. Extracting and preparing the input data
    b. Editing input and correcting data errors
    c. Processing data
    d. Updating information
    e. Preparing output documents
    f. Retaining data

   The design should be kept at the logical level, and you can determine the data requirements.

MINICASE—CONTEXT INTERNATIONAL, LTD.

During your early analysis of the financial information needs at Context International, you discovered that the credit monitoring function was not working well throughout the firm, largely because credit personnel lacked access to required information. In designing the new system, you will conceptualize a uniform credit monitoring capability to be used at each subsidiary. You have identified the following important systems functions as needing support:

1. The ability to check a customer's credit status, including credit limit per order, total dollar credit line, outstanding receivables, and credit rating
2. The ability to change a customer's credit rating
3. The ability to review a customer's credit and payment history
4. The ability to verify that a pending order should be allowed, based on the current receivables, order limit, and credit line limits established for the customer

5. The ability to report on delinquent accounts
6. The ability to produce a credit report that shows outstanding receivables by customer for intervals of 30, 60, 90, and 120+ days

As part of your design work, you must:

- Present the design in overview fashion (for example, prepare a VTOC diagram)
- Design CRT screen formats for all online transactions (include a transaction menu screen, as well as error message screens)
- Design report formats, as required
- Design the systems flow, indicating the required inputs, processes, and outputs (be sure to indicate any editing that you believe is required)
- Prepare your results in a form suitable to the needs of a review team that will be involved in a structured walkthrough of your design

9

File and Data Base
Organization and Design

CHAPTER OBJECTIVES

When you finish this chapter you will be able to—

- *Discuss the management information system (MIS) concept and the different management levels served by a functioning MIS.*
- *Identify the resource requirements of a computer-based MIS.*
- *Identify and discuss the three points of emphasis in designing an information base.*
- *Identify and discuss conditions that could create problems in designing an information base.*
- *Define the different methods of organizing and accessing files.*
- *Identify and discuss three data base models.*
- *Describe the important functions performed by a data base management system (DBMS).*

Central to the whole task of systems design is the organization of data into an effective and efficient information base. This is particularly true in contemporary computer-based business systems whose purpose is to make information available for decision making. In this chapter, we will explore information base organization and design by attending to four major discussion topics:

1. Management information systems
2. General design considerations
3. File organization and design
4. Data base organization and design

These four issues illustrate the theory and practice of designing a system that effectively manages information. The term information base is a general one used to describe any body of stored information. There are two primary kinds of information base, the *data file* and the *data base*. Recognizing that there are times when the two terms are used as if they were interchangeable, it is probably best to distinguish them as follows:

- Data file—a body of information which is organized for a particular application or limited series of applications, for example, a payroll file.

- Data base—a comprehensive body of information which may be selectively organized in a number of different ways to meet a broad range of different needs, such as an employee data base in which payroll data is only one aspect of the total; a data base is under the management control of a data base management system (DBMS).

MANAGEMENT INFORMATION SYSTEM (MIS)

Management Information Systems (MIS) are an attempt to relate information to its intended management use. In other words, MIS development is an attempt to organize and classify information based on what level of management will use it.

You will recall, from the prior discussion of problem complexity in chapter 4, that the following levels of systems were indicated:

1. Business Activity Subsystem
2. Business Function Subsystem
3. Business Operation Subsystem
4. Business System

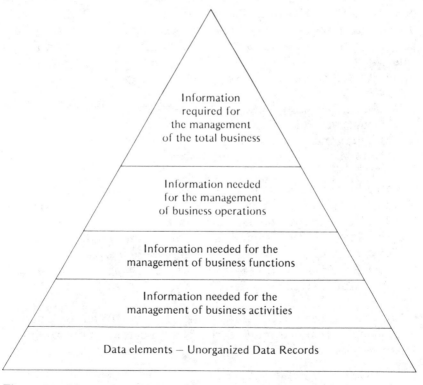

Information required for the management of the total business

Information needed for the management of business operations

Information needed for the management of business functions

Information needed for the management of business activities

Data elements — Unorganized Data Records

Figure 9-1   Management Information System—an attempt to relate information to its level of management use

The information required varies from one systems level to another. The information required to manage a systems activity differs greatly from that required to manage a total business, both in the degree of detail and comprehensiveness. To manage a business activity requires very detailed information about the specific activity. To manage a total business requires very concise summary information of the total business process.

Figure 9-1 is an MIS as a pyramid of information. At the very bottom of the pyramid, we find a base of data elements, or discrete pieces of unorganized information. As we proceed up the pyramid we find that information is being increasingly summarized and also that it is more comprehensive of the total business process. Figure 9-2 illustrates this inverse relationship between MIS detail and comprehensiveness perhaps more clearly.

The hourglass-like illustration shown in figure 9-2 underscores the importance of the base of data elements for decision making at all levels of management. That is, while decision making at the uppermost management levels requires a relatively small amount of summary information, the basis for that summary management information is really the broad base of data elements which supports it. Let us briefly examine each of the four management levels shown in the previous illustration.

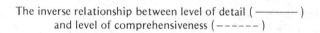

The inverse relationship between level of detail (————)
and level of comprehensiveness (−−−−−−)

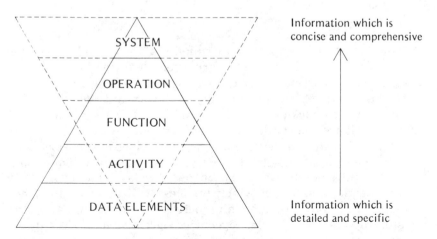

Figure 9-2   Management Information System—the inverse relationship between level of detail and level of comprehensiveness

Activity Management

While personnel in charge of specific business activities are frequently classified as supervisory and not management, the distinction is not always as clear as job titles would imply. Since there is no question but that some information-based decision making is required at this level, it is the first level of the MIS.

At this MIS level we are referring to information which will assist in the control of processing day-to-day business transactions. Some examples are information about the number of transactions being processed on a daily or monthly basis, information regarding the performance of specific personnel, and the existence of work backlogs. All provide vital evidence upon which management decisions can be made.

Functional Management

The management of a functional unit, such as a payroll department, can be a fairly complex job. Generally speaking, the head of a functional unit is not only responsible for workload processed by the unit, but also for management of the unit in a more comprehensive sense, which includes planning, budgeting, staffing, and reporting. In order to effectively perform these functions, the manager must have information regarding such things as workload trends, operating costs, personnel requirements, and current workloads.

It would be very cumbersome for the unit head to deal with the information used at the activity level, particularly because that information is not organized in a way which lends itself to decision making at this higher level. An effective management information system provides a summary of information for all activities within the business unit.

Operations Management

A major business operation is at roughly a vice-presidential level within an institution. The information required for effective decision-making relates and compares what is happening at the functional level. For example, it may be important to determine the extent to which backlogged orders are related to either strong demand for all products or demand for a few "hot" items.

Information at this level must be very concise and very comprehensive if it is to result in effective management decision making. This information will be used by executive management to determine the manner in which business will be conducted. For example, noting the relationships between production, sales, and existing inventories of goods, executives could effect a shift in the production of goods and in the discount pricing of those already produced.

Total Business Management

Information from the top of the pyramid is used to make decisions regarding what a business institution will be like ten or fifteen years into the future. For example, whether to implement a comprehensive multi-million dollar program to further develop and improve the existing company product lines (such as luxury automobiles, whale-bone corsets, and bottle cap openers) as opposed to investigating new product development is a decision of obvious importance to an institution's future. The need for comprehensive and concise information upon which these kinds of decisions can be made is of crucial importance to an organization.

Computer-Based MIS

There are those who argue that management information systems exist with or without computerization. Certainly, a computer is not required to structure information; but it should be quickly added that the comprehensive structuring of information within a complex modern organization is most reasonably accomplished through the use of computers.

A large resource commitment is needed to support an effective computer-based MIS. It requires at least the following:

Requirement	*Resources Needed to Support Requirement*
1. Broad base of business information	Extensive computerization of the business system
2. Data base management system	A computer with considerable main memory
	Extensive direct-access computer storage
	Higher level programming languages
	Sophisticated computer software
	High level of technical expertise among data processing personnel
3. Management access	Data communications capability
	Remote computer terminals
	Supportive computer software

The rather obvious extent of resources has caused many middle-sized and smaller organizations to shy away from this sort of "total" systems concepts in the past. However, a number of important occurrences in the past few years have caused many to reassess the reasonableness of computer-based MIS.

In recent years the development of data base management systems by computer vendors and independent computer software companies has stirred considerable interest in organizations that do not possess the depths of technical expertise required to develop their own. In most instances these software "packages" can be purchased or leased for far less than the cost of development. Provision is also frequently made to enable the software to interface with remote computer terminals.

Other factors influencing attitudes towards MIS include generally lower prices which have been forthcoming for computer hardware—including main memory, terminals, and mass storage devices.

Adding into this mix of ideas the very real fact that business is becoming more complex, more varied, and more subject to rapid change, we find that a situation exists in which computer-based MIS is becoming more of an advantage today than ever before.

MIS Application

The potential benefits of the MIS approach will, it is hoped, become more apparent as one examines the business structure of an organization—in this instance, the Typical Manufacturing Company. For explanation purposes, the example in figure 9-3 is somewhat simplified.

Simply noting the kinds of information required to manage Typical Manufacturing does not dramatize the extent to which management decisions require an interdependent treatment of that information. For example, the physical process of completing a sale is a somewhat isolated activity, provided one is "given" what products are available and at what price—and that is precisely the point to be made. The management decision as to what products should be sold has a great deal to do with marketing studies, as well as with the research and development activities within the firm.

The decision to build a new manufacturing facility may cut even deeper across the lines of available information. Questions such as the extent and nature of current facilities use must be related to production planning, marketing, and sales trend information.

Ultimately, decisions affecting the firm's goals or purposes will require an interpretation of information affecting most, if not all, aspects of the firm. To attempt these decisions without fully knowing the relationships among marketing, manufacturing, sales, and other

MIS for the Typical Manufacturing Company

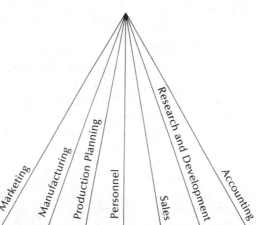

Figure 9-3   MIS for the Typical Manufacturing Company

vital information is to base a decision on limited understanding of the problem. The MIS concept is to present information to management in a manner which recognizes the need for summary cross-functional decision making.

GENERAL DESIGN CONSIDERATIONS

It is worth reemphasizing the point that a data base is the repository for data. The data base management system is a complex package of computer software that controls access to the data base, as well as to other important data processing functions.

Normally, when designing the information base, one distinguishes between logical and physical design. In logical design, we establish how the information organization will look to the systems analyst. Physical design refers to the actual machine location of the data. In this chapter, we will deal primarily with the logical design of the information base, with occasional references to aspects of physical design.

In attempting to design an information base which will effectively meet the needs of an organization, it is necessary to define the information base according to its:

1. Sources
2. Characteristics
3. Purposes

In so doing, it becomes possible to view information as part of the dynamic systems process.

Identifying Sources of Information

Someone or something has responsibility for entering or preparing the information for the information base. By noting these sources of information, the analyst is better able to control the flow of information into the information base. When we consider the important management decisions which will depend upon this information base, the importance of controlling its quality at all points of entry becomes apparent.

At times there is a single external source which has responsibility for preparing and entering particular information into the information base. Under this condition control of input quality is simplified. For example, the payroll department may be responsible for entering the number of hours worked for all employees. Control over the entry of this information is simplified because we are probably dealing with a few people, stationed in a single department. Also, in the event of errors, these personnel will probably also take the necessary corrective action.

In many instances, however, information may come from several sources, and the analyst must take care to design the system in a way which clearly identifies the source. For example, newly produced goods could be reported from several manufacturing sites, each of which has its own management hierarchy. If these plants also produce many of the same products, it will be impossible for the system to pinpoint the source of particular information unless it has been designed to do so.

Finally, it is possible that the source of information is an internal systems process. For example, in a payroll system the field of information called year-to-date wages is something which may be computed, updated, and maintained automatically by a programmed routine.

However simple the process of identifying the source of information may appear, the importance of doing it as one attempts to establish an information base cannot be overstressed.

If there is to be some assurance that information will be entered into the system both accurately and reliably, then the analyst should examine the proposed source to determine if it is the best possible source for the required information. Some of the questions which should be posed by the analyst regarding this matter include:

1. Is the information of little real interest and importance to the source?

2. Is the intended use of the information potentially threatening to the source?

3. Does the required preparation of input to the information base place an added workload burden on people at the source?

4. Does the time lag between entry of the information and the updating of the information base encourage the maintenance of an external information base?

If the answer to any of these questions is "yes," then there are potential problems related to the use of the proposed information source. Before the analyst can place full confidence in the information base, some additional work must be done to reduce or eliminate these factors which, if allowed to continue, might encourage carelessness, side-stepping, and even dishonesty in the preparation of input.

Identifying Information Base Characteristics

Information base characteristics may determine how the information should be organized and accessed. Consequently, it is important to define information and information bases according to the following characteristics:

- Size
- Variability
- Volatility
- Activity

The size of an information base will have obvious implications for selecting the media upon which the information will be stored, particularly when there are existing equipment limitations to be considered. The total size may also affect decisions regarding how information will be organized and processed. For example, the sorting of massive amounts of information will probably be avoided when possible.

The size of a particular element of information will also have consequences worth noting. For example, the size of a particular element of information called "state-of-residence" may be sufficiently large to require internal coding, such as a number 1–50 instead of the alphabetic state name.

It is important to distinguish between information which is constant (always the same) and that which is variable (subject to change). For example, college grades represent a variable field of information in that a student might receive a letter grade of A – F depending upon academic performance. However, in many grading systems, the letter grade of A will also have a constant value of 4.0 quality points, a B will equal 3.0 quality points, a C = 2.0 quality points, and a D = 1.0 quality points.

Generally speaking, variable information must be stored within the information base, but constant information will not. Constant information will be applied to the variable information base when it is needed. For example, a computer program which contains the constant quality point values and the necessary logic may produce from a student information base a final student grade report which shows both grades and quality points earned by the student.

Volatility refers to how often information will be added, deleted, or changed over a given period of time. When there are many changes to the information base we say that it is volatile. When there are relatively few changes to information, we say that the information base is static. Some information fields within the base may be more volatile than others, and it is a good idea to know which ones they are. For example, the particular students attending a college may show little volatility, in terms of changing their names once they are enrolled. However, there will be great volatility in terms of the particular courses which they are enrolled in over the span of two to five years.

Closely related to the characteristic of volatility is that of activity. Activity refers to the percentage of the total information base which is being referenced for a given period of time. This does not necessarily mean that the information must be changed to be active; it could be that it is only being read. Furthermore, it could be that an information base was simultaneously characterized as highly volatile and low in activity. For example, a massive information base in which thousands of changes were occurring each month; but where these changes were being made to only 5 percent of the total information would constitute just such a case.

The relationship between information base characteristics and the organization and accessing methods will be discussed later in this chapter.

Identifying the Purposes of Information

Identifying who will use information and how they will use it has important implications for how information is stored and where. For example, if several fields of information are required by the same business unit as part of a single decision-making process, then it would seem sensible to store that information either physically together or in a way which enables one to access it as if it were physically together. In any event, we would want to avoid locating the information in a way which makes it difficult or impossible to retrieve it as a part of the same report or CRT display.

There are also times when information is important to more than one individual or department. At such times it is necessary to organize the information base in a way which supports common or shared usage.

As we shall see in the following sections, there are numerous methods for organizing and accessing information. From this menu of possibilities it is possible to find alternatives which will be consistent with the purpose or intended uses of that information.

In those sections you will be introduced to several major concepts dealing with the organization and accessing of files and data base management.

For a professional systems analyst, understanding these basic concepts is important—particularly because they will affect the analyst's ability to make wise judgements regarding the information organization and accessing methods to be employed in a given problem situation.

Data Records

Before we can discuss methods of file and data base organization, some understanding of data records is necessary. A data record is a group of associated data elements which refer to some unit, such as a customer, student, or inventory item. The data record may contain all of the available information regarding the unit in which case it is called a master record; or it may contain only a particular kind of information about the unit in which case it is called a detail record. Following is an example of the layout for a master personnel record.

Master Personnel Record Data

A. Personal Data
   1. Social Security number
   2. Employee name
   3. Home address
   4. Spouse's name
   5. Home telephone number
   6. Office room/building
   7. Office telephone number
   8. Birthdate

B. Appointment Data
1. Date of first appointment
2. Current job title
3. Department
4. Current salary
5. Appointment date
6. Supervisor's name
7. Supervisor's title

C. Employee Benefit Data (for each benefit type)
1. Benefit type (for example, retirement, health, or dental)
2. Firm contribution
3. Employee contribution
4. Vesting date
5. Provider

D. Employee Credentials
1. Highest degree awarded
2. Degree-granting institution
3. College grade-point average

As part of the design for a new system the analyst must determine how the various data elements (also called *fields*) will be organized into data records. This is referred to as preparing record layouts. There are a number of guidelines to be followed in preparing record layouts.

Information which is needed to perform important processes should be contained in the same record whenever possible. The primary reason for this is that computers frequently access and use information at the record level. If these data elements were contained in a single data record, it would be possible to access all of the relevant information in the amount of time which it takes the computer to access or read a single record. If the same information is placed into six or seven records, it may take longer to read the records and to keep track of which record is being read at any given time.

A second consideration in performing a record layout is to place the most critical data elements to the front of the record. If the name of an individual is the data element which most clearly identifies the record, then it should be placed first in the record. This practice simplifies manual reading of record lists which are obtained from the computer.

Related data elements should, whenever possible, be placed together within the data record. This practice, once again, eases working with record listings.

Data files and data bases contain series of data records which can be organized and accessed by key identifiers.

FILE ORGANIZATION

Even though one may work only with data base management systems, it is important to have some basic understanding of conventional file organization and access methods. Most data base management systems logically employ variations of these methods.

In this section, we will discuss:

- File/record keys
- Sequential files
- Random files
- Index sequential files
- Unordered files
- Lists
- Rings
- Inverted lists

File/Record Keys

Two kinds of file keys will be discussed throughout the remainder of this chapter, namely, "primary" and "secondary" keys.

Primary keys uniquely identify data records. For example, in an accounts receivable system, the primary key to a customer's accounts receivable record might be a unique customer number. At this point, it is worth noting the difference between "logical" and "physical" keys. The logical key is a particular data element (for example, a customer number), whereas the physical key (also called the "actual" key) identifies the physical location or address of a data record within the computer's memory.

Secondary keys do not identify a unique record. Rather, they link many records that share a common attribute. For example, let us assume that, if a customer failed to pay his or her bill after 120 days, we would flag the account as being delinquent. If we specified the delinquent indicator as a secondary key, all delinquent accounts would be linked together, and all nondelinquent accounts would also be linked together. As we shall see in later discussions, a record can have many secondary keys.

Sequential Files

A sequential file is a series of data records which are organized and accessed according to some predetermined sequence. For example, a sequential file of all employees in your firm might be arranged either alphabetically, starting with the last name; or by Social Security number starting with the low order number. While the sequential file has some flexibility in that it may be placed in any sequence for any data element through proper sorting, it can only be accessed on the basis of that sequence. Figure 9-4 illustrates the sequential file method.

There are some important characteristics associated with the use of sequential files which are important to systems design:

1. Sequential files may be stored on a wide variety of media, including punched cards, tapes, disk, and magnetic drum.

2. Sequential files may be read (input file) or created (output file).

3. Information is accessed on a sequential file by starting at the beginning and continuing to search the file until the right key (for example, person's name) is found.

4. Sequential file records can be block read. That is, a single computer reading can be made of many data records. This means that the need for large master records is not as critical for sequential files.

5. Sequential files are normally updated by reading the old file and creating a new file which contains relevant old data plus the update information.

Sequential files have certain advantages when the file activity level is high. When all or most of the file information must be accessed as part of an average job, sequential ordering makes a great deal of sense.

Access Order	Personnel Record File	
	Key	Other Data
1	Alpha, Joe	Data
2	Alpiz, John	Data
3	Beta, Sue	Data
4	Boza, Jill	Data
5	Charlie, John	Data

Figure 9-4 Sequential file organization

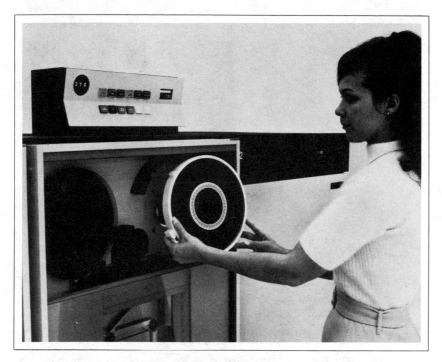

Magnetic tape unit for reading and writing sequential files

A second advantage is that there is little or no wasted space within the file itself. Records appear one after another with no blank data areas until the end of the file is reached.

When the file is characterized by low activity and high volatility, sequential organization becomes a handicap. The continual reading and writing of masses of information for frequent changes to a small portion of the file information is cumbersome and costly.

One of the very real limitations of sequential files is the inability to go in and specifically access one record without also reading all of the records which precede it. In most of the other file organization methods which we will discuss, it is possible to directly access a specific data record.

Random Files

Random file organization is designed to permit speedy access to any given record within a file without first reading all of the records which precede it. For this reason it is one of several file organization methods commonly referred to as *direct file access*. Information which is to be accessed directly must be placed on a direct access

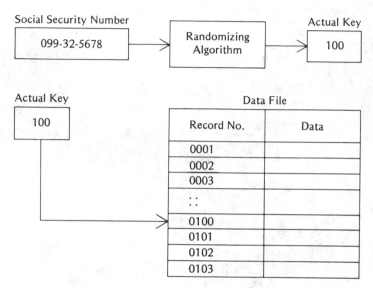

Figure 9-5 Using a randomizing algorithm to convert a Social Security number to a file location

storage device, such as magnetic disk or magnetic drum memory. Magnetic tape, punched cards, and punched paper tape cannot be used as the storage media.

The word *random* is used to indicate that file records are not ordered according to any of the data elements which they contain, such as name or social security number. Rather, they are ordered as a result of scrambling or randomizing one or more of the data elements to determine the exact location (ACTUAL KEY) of the record in the file.

Figure 9-5 shows how a randomizing routine or algorithm might convert a social security number to a file location (in this case it is record number 100 in the file). Once that file location is obtained, the specific data record associated with that key can be accessed.

While random file organization allows direct accessing of a particular record, it does so at the expense of maintaining any sequence to the file. Consequently, extensive sorting of the file is required before output can be obtained in a desired sequence.

At times the randomizing algorithm will yield two or more identical actual keys for different records. These records are called synonyms and must be given different storage areas. Only one record will be placed in the main or base data file. All synonyms will be placed into a file overflow which is linked to the base file, as shown in figure 9-6.

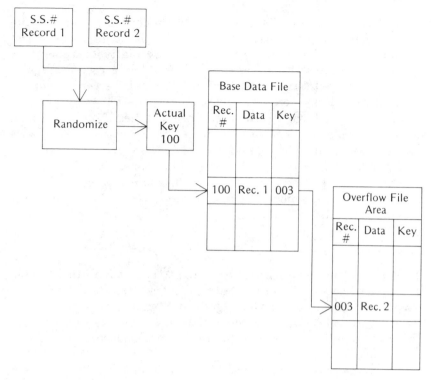

Figure 9-6 Overflow file linked to base data file

Without going into greater detail regarding structural aspects of random files, we want you to remember that:

1. It is possible to have two or more records with the same actual key (called synonyms).
2. All synonyms are placed in a file overflow area which is linked to the base area record by a second actual key which we have called the link key.
3. As the number of linkages increases with multiple synonyms, so does the "average" number of accesses required to read a data record.
4. When a record is deleted from a random file it does not always mean that additional space will be made available, since
   a) it is only available to a synonym, and
   b) all linkages must be maintained or records may be lost.
5. Whenever it is necessary to obtain information in some desired sequence, it will be necessary to sort the random file to that sequential order.

The advantages and disadvantages of random files must be considered against the characteristics of the file. Random file organization works best when the information base is relatively stable since this means there is little need to add and delete records. It also works well for situations in which only a small percentage of the total file records are accessed during an average job run.

Difficulties will arise in the use of random files if there is a high level of file volatility since the number of links between synonyms will increase along with the number of accesses required to read a record. When these linkages become numerous enough to affect accessing speed, it is necessary to perform maintenance of the file. This maintenance removes all deleted records from the file and thus lowers the number of linkages.

Index Sequential Files

Index sequential file organization is an attempt to maintain the file in a desired sequential order, but also to permit random access to any particular record. This file organization will allow speedy direct accessing and also eliminate some of the need for extensive sorting. This is accomplished through the use of three separate files:

1. A course table (sometimes called the "rough" table)
2. A fine table
3. A data file

The sequential accessing of the file is possible because both the course and fine tables, which are small files, are sequenced by the desired data element (such as social security number). The direct accessing of the file is possible because an actual key is associated with each fine table record. Figure 9-7 presents an example of how index sequential files work.

The course table contains a limited number of entries which cover the full range of record possibilities. For example, the first entry of

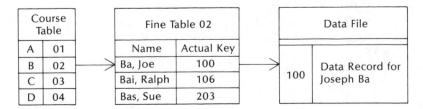

Figure 9-7   An index sequential file

the table may cover all last names beginning with the letter A. The second covers all last names beginning with B and so on until the last entry which covers all last names beginning with the letter Z. In addition, the course table points to a fine table, which has a complete list of all data records within that range and the actual key of the record.

Thus, by rapidly searching the very small course table one is able to go to a limited list of sequentially ordered entries. A sequential search of those entries will direct one to the actual file location of the data record.

Also, the entire file can be read sequentially by simply following the sequence of the course and fine tables.

Because only the relatively small course and fine tables have to be maintained in sequential order, maintenance is relatively easy. However, as with random files, a high amount of volatility within the data file can lead to excessive linking of records within one data file. An extensive maintenance of the data file is required to eliminate these extensive links, which will otherwise slow down the average file accessing time.

Unordered Files

As its name implies, an unordered file is simply a series of records placed one after another as if in a large tank. There is no unique identifier associated with an unordered file and so there is no way to access a specific record within the file. The records can be accessed serially, but they are in no special order.

One important characteristic of an unordered file is that it is unaffected by volatility. As records are deleted from the file, the space becomes available for the next add to the file. Available space is located by file address, and it requires only one access to add information to a file.

While the use of unordered files is in many ways a step back from the sophistication of random and index sequential file structures, unordered files are frequently used in data base management systems. This use stems from the fact that, although the unordered file record has no key identifier of its own, the record can be accessed through keys from other associated files and through the list and ring structures explained in the following sections.

Lists

Recall from our earlier example of a random file overflow record that two records may be linked by having the primary key to the second

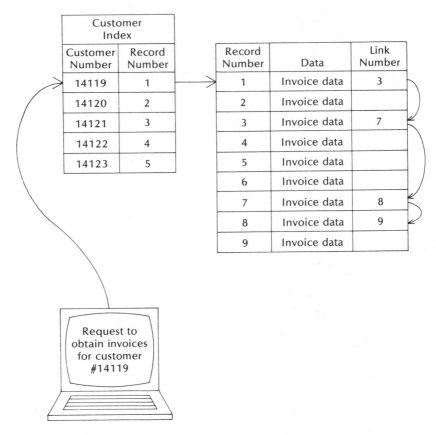

Figure 9-8 Example of a list

record stored in the first record. When this linking is used for a whole series of records, the result is called a list.

Figure 9-8 illustrates how, in an unordered file of customer invoice records, access is first made through a customer index to obtain the first invoice record. Each succeeding record is accessed through a link to its predecessor.

Rings

Ring organization is an extension of the simple list concept. Each record in the ring contains pointers to both the preceding and following records. For example, let us assume that we have stored a customer's invoice records by billing date, using a ring structure, as in figure 9-9. Because the records are in ring form, we can more readily retrieve them in either last-in-first-out or first-in-first-out order.

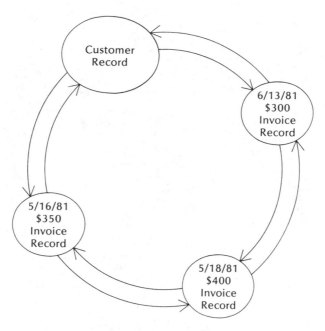

Figure 9-9 A ring structure for storing a customer's invoice
records by billing date

Using a more complex example, let us also say that each customer
cash payment received is stored in a ring around the invoice it
applies to, as in figure 9-10. This figure shows that "closing the loop"
has another advantage. That is, we can momentarily leave one ring
(for example, invoices) to obtain related information (for example,
cash payments), then return to and continue in the first ring.

Inverted Lists

In the earlier discussion on secondary keys, it was pointed out that
it may be desirable to link records that have common characteristics. An inverted list is simply an extension of the list concept, in
that a single record may be linked to many lists, based on common
characteristics instead of common primary keys. For example, as
figure 9-11 shows, the personnel record for a staff member named
Lopes may be linked to other staff records, based on sex, eye color,
or other common characteristics.

Note that, in this approach, all the record key linkages are maintained in the characteristic index rather than in the records themselves. The reason is obvious—individual records may be linked for
virtually all characteristics, and we would not want to provide key
space for all possible lists in each data record.

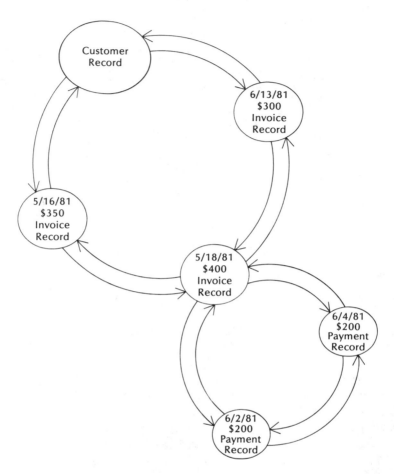

Figure 9-10   A ring structure for storing each customer payment around the invoice being paid

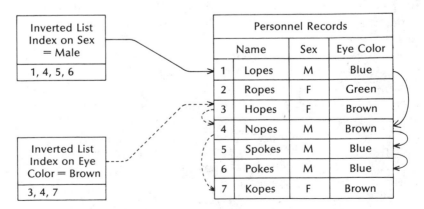

Figure 9-11   An inverted list

DATA BASE DESIGN

Data base design is a complex topic, and any in-depth treatment of the subject exceeds the objectives of this book. However, it is possible to provide a conceptual overview of the subject that will prove useful in supporting future, more extensive study.[1]

In this section, some attention will be given to each of the following:

- Data base management systems
- Top-down schema design
- Data base models

Data Base Management System

Let us clarify at the outset the difference between a data base, which is the repository of data, and the DBMS, which is the vehicle for controlling access to and modifying the data base. Because of this intermediary role, the DBMS must effectively interface with the physical data base, user application programs, data processing requirements, and essential systems software, such as software that supports data communications. Also, the DBMS frequently contains a number of features that address specific issues of concern to the systems designer, such as systems security, recovery, and audit trail capabilities. Let us look at several of these features and interfaces in greater detail.

Interface with the physical data base   To effectively interface with the physical data base, the DBMS must have available a complete description of the data and their organization. This global description of the data base is commonly called a "schema." It is also possible to have subset descriptions for portions of the data base, which are called "subschemas." An obvious use of the subschema approach

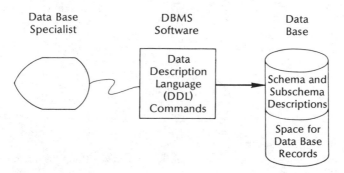

Figure 9-12   Building schema and subschema descriptions

is to limit or restrict program or user access to only those portions of the data base that are needed and that access is authorized for. Remember that access to the global schema means that one can find the location of all data. In establishing a data base, one of the first tasks is to develop schema and subschema definitions, using a special data description language (DDL) provided as part of the DBMS software, as in figure 9-12.

Interface with Data Processing and User Application Program Needs Using a higher-level programming language like COBOL and a conventional file structure such as a sequential file, the programmer can easily control the logic commands required to retrieve or modify data. In a complex data structure such as a data base organization, which employs numerous methodologies (for example, lists, rings, and so on), the DBMS provides an effective interface between the traditional program and the physical data base. This interface is called the data manipulation language (DML). The DML is a series of special data base instructions that the DBMS software understands, and they interact with the schema and subschemas in controlling the information flow to and from the data base. Figure 9-13 shows a simplified view of this process in an online environment.

Additional DBMS Features A wide range of other features may be, but are not always, either a part of the DBMS or interfaced with it. The systems designer will want to know which of these features are

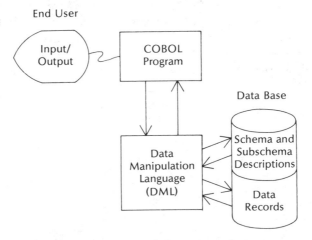

Figure 9-13 Use of the data manipulation language (DML) by an application program

available with any particular DBMS, because they have broad implications. They include:

- Data base security features
- Data base recovery capability
- Audit trail capability for online transactions
- Data communications interfaces
- Query language and report writer capabilities
- Data dictionary capabilities
- Various data base utilities that can be used to analyze the efficiency of the data organization, load and unload the data base, facilitate the reorganization of the data base as it is expanded, and so on

Top-down Schema Design

Unfortunately, many designers make the same mistake in data base design that has traditionally plagued functional design efforts. That is, they begin at a very detailed level (for example, defining data elements) and then attempt to reconstruct the more global, integrated picture.

As with functional design, it is highly recommended that the data base design begin with a comprehensive overview. For example, if we know that our data base will contain information about people,

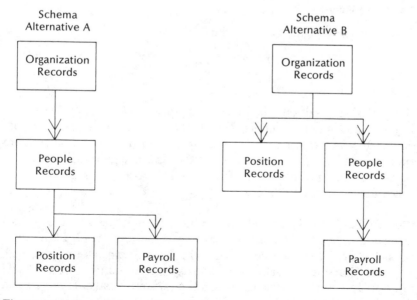

Figure 9-14 Two top-down schema designs

organizational structure (for example, departments), position titles, and payroll amounts, it becomes important to first establish their broad interrelationships for the particular system being designed. Figure 9-14 presents two possible top-down schema designs. Which alternative is more appropriate?

In alternative A, both payroll and position records belong to people who, in turn, belong to particular organizational units. Alternative B differs in that both positions and people belong to organizations. Either of these alternatives may be appropriate or inappropriate for a particular system. Once it has been established which is more appropriate, one can work down to more specific levels of detailed description, such as which data elements should be placed in which records.

At this point, it is worth noting a second potential pitfall in data base schema design, namely, the occasional tendency to make the schema overly complex. Remember that any request to access or modify data will have to "navigate" through the schema, exploring all the various linkages and making many individual record accesses until the desired access or change has been completed. As fast as computers operate, an overly complex (or poorly designed) schema can result in significant processing delays. As with most aspects of design, simplicity in schema design is generally preferred, provided that the basic system requirements can be fulfilled.

Data Base Models

There are three basic types of data base models:

- Hierarchical (or tree structure)
- Network
- Relational

Many DBMS products on the market today can be grouped under one of these generic models. Naturally, one vendor's use of a particular concept or model (for example, hierarchical) differs from another vendor's. However, these differences are generally practical variations on a fundamental, conceptual theme. Let us briefly introduce each of these models.

Hierarchical Model   In the hierarchical model, the schema design reflects a tree-like structure in which some data records (or data elements) are subordinated to others in the way they can be accessed. This relationship among data records is commonly called a "set" relationship, with one record called the owner (or parent) record and the others called member (or child) records.

Figure 9-15, a simplified schema for a personnel/payroll system,

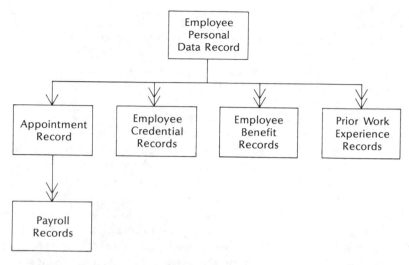

Figure 9-15 Simplified personnel/payroll schema (hierarchical)

presents a hierarchical relationship among employee, appointment, credentials, benefits, prior work experience, and payroll payments. The single arrowhead in this figure represents a one-to-one relationship between owner and member records (for example, any employee is permitted one and only one appointment in the schema). The double arrowhead represents a one-to-many relationship (for example, an employee can have many prior work experience records, perhaps one for each previously held job).

In a hierarchical model, a member always has one and only one owner, whereas an owner can have many members. In other words, it may be necessary to duplicate some records if two owners need to be superordinate to the same member record information. For

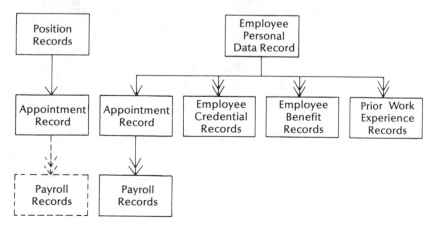

Figure 9-16 Hierarchical schema with redundant data records

example, we may wish to permit both employees and position records to own job appointments, because appointments are to a position and also belong to employees. In the hierarchical model shown in figure 9-15, it would be necessary to duplicate the appointment data, as shown in figure 9-16. If we had wanted to maintain a relationship between positions and the amount of money expended on them, we would also have to duplicate the payroll records, as indicated by the dotted lines in the figure.

Natural hierarchical schema designs can become quite complex and extend down to several levels of hierarchy, but the basic principles outlined above remain essentially the same.

Network Model   As its name implies, the network model is not constrained by the single-owner rule of the hierarchical model. In our description of this model, we continue to employ the terms "set," "owner," and "member." Using the previous example, it is now permissible to have both employees and positions linked to appointments, as shown in figure 9-17. From this figure we can determine that, given access to the employee personal data record, it will be possible to access the appointment record and, from there, either the position or the payroll records. Given access to the position records, we could likewise network to any of the other records defined by the schema.

Because of its capacity for dealing with the multiple-owner situation, the network model generally creates less duplication (data redundancy) and more opportunities for interrelationships among data records than the hierarchical model does. However, as you increase the complexity of the possible interrelationships, the pro-

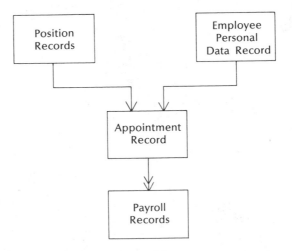

Figure 9-17   A network schema

gramming logic required to implement the design flows less readily. Note that data redundancy is not inherently "bad." When using the network model, sometimes it may be appropriate or even necessary to have redundant data by design.

Relational Model   There appears to be at least some support for the notion that the evolution of science has been characterized by a gradual move from the conceptually complex to the conceptually simple. For example, contrast the design of an internal combustion engine with that of a jet engine, or contrast early electromechanical devices with their modern electronic counterparts.

Many believe that we are on the verge of a similar breakthrough in DBMS design because of developments that have led to a relational data base model.

In the relational model, there is no hierarchy, and thus it is commonly called a "flat" file structure. Also, there is no complex network that one needs to navigate through. Rather, there is simply a series of data element relationships that are established in table-like form. Data elements that are related in this way are called a "tuple." For example, there may be a tuple called "employee" that consists of the following related data elements:

- Employee number*
- Employee name
- Employee address
- Employee telephone number

$\left.\begin{array}{l} \\ \\ \\ \\ \end{array}\right\}$ = "Employee" tuple

*Key data item

The concept is truly a simple one—namely, describe all the important data relationships in your schema. If you require new relationships, define them and update your schema. There is no presumed hierarchy that you have to navigate through.

Although the relational data base model has a straightforward conceptual approach, the main challenges it poses pertain to efficiency. For example, examine the following partially described tuples for a student information system:

- Bill (student identification number, last name, first name, street, city, state, zip code, amount billed, previous payments, and so on)

- Grades (student identification number, last name, first name, street, city, state, zip code, grade-point average, and so on)

- Awards (student identification number, last name, first name, street, city, state, zip code, total award amount, and so on)

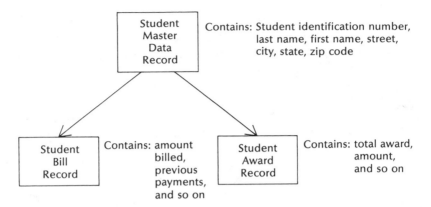

Figure 9-18   A hierarchical schema description

Although the above descriptions overstate the case somewhat, the point is a significant one, namely, that key data elements may need excessive restatement. Contrast this situation with the hierarchical schema description shown in Figure 9-18.

Yet the problems in relational data base efficiency are not permanent obstacles. They can be overcome, and much of the DBMS development work of the 1980s will focus on developing increasingly efficient relational data base structures.

CHAPTER SUMMARY

Establishing a systems information base is a central issue in systems design. There are two primary kinds of information bases that the analyst must be knowledgeable about: the data file and the data base.

The management information system (MIS) is an attempt to relate information to the intended management use of that information. Thus, through MIS development, it is possible to organize and classify information based on the level of management that will use the information. Computerization would, in most respects, appear to be a logical ingredient. The resource requirement is considerable for developing and operating an effective MIS and remains a major obstacle to its widespread use.

A number of important general design considerations warrant the systems analyst's careful attention. In attempting to design an information base that will effectively meet an organization's needs, the analyst must define the information base according to (1) sources of information; (2) characteristics of the base, including its size, variability, volatility, and activity; and (3) purposes the information will serve with respect to security, recovery, data communications, query, and reporting functions.

Many designers make the same mistake in data base design that has plagued functional design. That is, they concentrate on a very detailed level (for example, the data element level) before obtaining the global, integrated picture of the design. The advantages of top-down data base design are as significant as those for top-down functional systems design.

The primary data base models are (1) hierarchical (or tree structure), (2) network, and (3) relational. There are many DBMS products on the market today. Although these products differ, they can be conceptually placed under one of the above model types. The evolution of DBMS concepts is most assuredly a major trend in the data processing industry. Many obstacles to the practical use of DBMS technology by a wide range of organizations, big and small, have already been eliminated. At its current rate of development, DBMS technology will soon be an option for most systems development efforts.

A number of file organization and access methods exist, including sequential, random, index sequential, and unordered files, as well as data structures such as rings, lists, and inverted lists. The characteristics of the information base directly relate to the effective use of any given data structure. The analyst must take care to select a file organization and accessing method that suits the characteristics of any particular information base.

To perform data base design effectively, the analyst requires some basic knowledge about (1) data base management systems, (2) top-down schema design, and (3) the three major data base models.

The DBMS is a system of sophisticated computer programs that interface user application programs with the physical data base. The DBMS also performs many of the other coordinating and operational functions.

WORD LIST

actual key—the physical file address or location for a data record

data base—a comprehensive body of unorganized information which may be selectively organized in a number of different ways to meet a broad range of different business needs

data base management system—a series of computer programs used to structure and control access to the data base.

data description language (DDL)—a language, also called the data definition language, provided as part of the DBMS software and used to build schema and subschema definitions

data file—a body of information which is organized to service a particular application or limited series of applications

data manipulation language (DML)—a series of special data base instructions, provided as part of the DBMS software, that interact with the schema and subschemas in controlling application programming access to the data base

data record—a group of associated data elements which refer to some unit such as a customer, student, or inventory item

hierarchical model—one of the three major data base models, in which the schema reflects a treelike structure between data records, some of which are subordinated to others

index sequential file—a method of organizing data records which permits both sequential and random access to those records

information activity—a measure of what percentage of the total information base is being referenced during a given period of time

information volatility—the frequency with which information base data will be added, deleted, or changed during a given period of time

inverted list—a data structure that maximizes the use of secondary keys by enabling a single record to be linked to many lists, based on common characteristics as opposed to primary keys

list—a data structure in which a series of records is linked in chain-like fashion, with each record containing the primary key of the succeeding record in the list

management information system (MIS)—a method of organizing and classifying information based upon the intended management use of that information

network model—one of the three major data base models, which allows complex relationships to be established between records and is not restricted to a given number of owners or members for any one record

primary key—the unique identifier for a given data record

random file—a series of data records which are ordered and directly accessed according to their physical location in the data file

relational model—one of the three major data base models, which uses a "flat" file structure as opposed to hierarchical relationships and is currently in its early, experimental stage of development

rings—an extension of the list concept, in that each record contains pointers to both the preceding and following records, including a return to the first record from the last record

schema—a description of the data base organization that itself resides as a separate area in the data base and is the map enabling the DBMS software to access, retrieve, and update data that are stored in the data base

secondary key—an identifier for a series of data records that share a common characteristic or attribute

sequential file—a series of data records which are organized and accessed according to some predetermined sequence

set relationship—a relationship between two data records in which one record is the owner of the other (member) record

tuple—a data element relationship, used in relational data base definitions

unordered files—a series of serially accessible data records placed one after the other but in no particular order

QUESTIONS AND PROBLEMS

1. Identify and discuss conditions which could result in design problems for each of the following situations:

    Situation 1
    C. Rollen Stone Co. has been interested in the integration of its personnel and payroll functions for several years. One of the compelling reasons cited by management for this new integrated system has been the need for better information regarding employee productivity and production costs by department.

    After an extensive review of the problem by the company senior systems analyst Jean Papas, a design team has begun its work. Papas has developed a revised time sheet which will be used to update a comprehensive employee data base.

    Situation 2
    Motown Manufacturing Corp. is a major manufacturer of household furnishings. A large aggressive sales force services a full range of retail and wholesale businesses. Because many small orders are frequently involved, sales personnel generally cover a large geographic area.

The company would like to know more about the kinds of customers it is servicing, as well as its own image with those customers. Accordingly, the company has designed a modified "order form" for use by sales personnel. The new form requires that the sales person ask the customer a number of questions regarding the customer's business and the customer's feelings about Motown.

It is anticipated that the new order form information will be added to the company's existing customer information base.

2. A new financial and accounting management information system has been developed at Mineral Mining Industries. The system uses data base management technology and has extensive online terminal features. The system affects seven different offices, each of which previously operated independent office-based systems.

Anthony Banfi, the lead analyst on the project, has decided to speed up the training and implementation process by instructing employees in only those aspects of the system that they must operate.

Based on your readings so far, what problems may result from Banfi's approach? What would you do differently, and why?

3. The order department's staff is overloaded with data entry work. You have been told that the accounting office has some free time and will "chip in" now and then. However, the accounting employees cannot be physically moved from their suite, and you will have to place a terminal in it. What don't you like about the proposed arrangement?

4. Every two weeks, you run paychecks for all employees from the payroll master file. Your manager prefers structuring it as a direct access file (for example, random or indexed sequential). If there were no other jobs using the payroll master file, what would you recommend, and why?

5. Given the following data relationships, diagram the ring structure(s) required to support them:

    • For an employee: one or more jobs worked on
    • For a job: one or more intervals of work by the above employee
    • For a job interval: one or more charges based on charge rate (for example, regular, overtime, or double time).

MINICASE—CONTEXT INTERNATIONAL, LTD.

The new financial system design will require a data base structure that supports the ability to track job costs for labor and materials consumed during the manufacturing process as well as an overhead rate assigned to the job. Using the network data base model, you will develop an overview schema that enables the user to:

1. Look at a particular job record and see the labor charges against that job for each individual, or the materials charges against that job for each inventory part number, or both kinds of charges
2. Look at a particular employee record and see all the specific jobs that he or she has charged time to
3. Look at the record for a particular inventory stock item and see which jobs particular parts have been charged to
4. See in summary form the labor, materials, and overhead costs for a particular job

As a follow-on to your general schema design, you should identify the kinds of data (for example, data elements) that you would expect to be included in each record type. You should also design appropriate input and output CRT screens that will be available to the user in each of the above instances.

NOTES

1. For a more complete study of DBMS, consult *Computer Data Base Organization* by James Martin (Englewood Cliffs, N.J.: Prentice-Hall, 1977).

10

Computer System Evaluation and Selection

CHAPTER OBJECTIVES

When you finish this chapter you will be able to—
- *Discuss the rationale for using a formalized approach to the evaluation and selection of computing equipment.*
- *Identify and discuss the formal steps in the evaluation and selection of computing equipment.*
- *Discuss the price/performance evaluation of equipment demonstrating knowledge of available pricing arrangements, methods of obtaining the "best" possible price, and alternative methods for evaluating performance.*
- *Discuss the systems-capability factor which is used in evaluation of equipment, demonstrating knowledge of both hardware and software capabilities.*
- *Discuss the systems-support factor used in equipment evaluations including field engineering, systems engineering, and education and training support services.*
- *Illustrate the relative importance of the major factors, subfactors, and categories used in the formal evaluation of equipment using a point-scale approach.*
- *Discuss and contrast the evaluation and selection of minicomputers and microcomputers versus general-purpose computers.*

Frequently, the development of new computer-based systems involves the purchase or rental of additional data processing equipment. This new equipment may be additional peripheral devices, such as disk and tape units or expanded memory. This kind of equipment enhancement is generally referred to as systems modification or systems reconfiguration. It is not regarded as a new system because the central processing unit (CPU), or computer mainframe, has not been altered or replaced.

When the computer mainframe is significantly modified or replaced, it has been traditionally assumed that system "upgrading" or "downgrading" has occurred and that a new computer system has been installed. Because of the increased modularity of today's computer systems and because the CPU itself is not always the best gauge of overall systems performance, the terms "upgrading" and "downgrading" have lost some significance. Nonetheless, for purposes of definition and in keeping with most computer manufacturers methods of systems classification, we will continue to say that a new computer mainframe is synonymous with a new computer system.

A third kind of equipment enhancement is the addition of special-purpose stand-alone data processing equipment. Programmable accounting machines, air-sampling devices, and traffic light regulators are examples of this kind of equipment. While some of the materials presented in this chapter could be applied to special purpose equipment, it is not our primary area of concern. Rather, the discussion focuses upon the evaluation and selection of new computers and computer reconfiguration projects.

Because the systems analyst is intimately involved in the design of the system, he or she is frequently called upon in the evaluation and selection process. If the analyst is to be effective in this role, additional knowledge of how to evaluate and select equipment is essential.

THE NEED FOR FORMAL EVALUATIONS

In the early years of data processing, many systems professionals relied heavily upon specifications or proposals prepared by computer equipment suppliers for the information upon which their final equipment decisions were made. Also, independent consultants and associates using similar equipment were primary sources of information. Lastly, prior success or failure with one computer model was a primary factor in "sticking with" or changing computer vendors when it came time to enhance one's computing capability.

While there is some value in each of the above approaches, methods of equipment evaluation and selection have changed in recent years. Today, most users of data processing equipment have taken the stand that the process should be formalized. Computer vendors should be called upon to demonstrate the capabilities of their equipment in meeting the workload needs of the user. There are several reasons for the shift to formal evaluation and selection.

It is unlikely that computer system performance can be accurately estimated from written specifications. Performance is the result of interaction between the hardware and software components of the system. The way in which hardware and software interact in getting the job processed must be judged after "live" demonstrations.

Second, a computer system may not work equally well for all types of workloads. Thus, the analyst must evaluate its performance in processing a specific kind of workload. The prior success or failure of friends and associates with particular kinds of computers may be meaningless if there is no similarity between workloads.

Finally, the development of new and different computer processing techniques by different computer vendors has made it difficult, if not impossible, to obtain comparative evaluations of competitive equipment without formal evaluation of these computers for specific kinds of workloads.

THE FORMAL EVALUATION AND SELECTION PROCEDURE

In formalizing the evaluation and selection process, a series of well-planned steps is developed and executed by the person or persons responsible for evaluation and selection whom we will call the computer user. A timetable of events must be developed which realistically reflects the amount of time which will be consumed by each step in the process. Gantt charts and/or the critical-path method, which will be discussed in the next chapter, are frequently used in constructing this timetable.

Because questions often arise the user must assign someone to manage the flow of information between the user and computer vendors who intend to propose their equipment. Although there are some variations, normally the formal procedure followed by the user includes:

1. Workload Analysis
2. Development of Equipment Specifications Proposal
3. Invitation or Request for Proposal (RFP)
4. Proposal Review and Evaluation
5. Equipment Demonstration and Evaluation
6. Final Evaluation and Selection Proposal
7. Vendor Notification and Debriefing
8. Contract Negotiation

Workload Analysis

The first step in the formal evaluation process is to analyze existing and projected workloads. Only in this way can the user begin to obtain a clear picture of true equipment needs. Methods for doing this are more fully discussed in appendix C. The primary objective, regardless of the method employed, is to develop a representative sample of one's workload, which can be tested on any proposed equipment. As mentioned previously, this representative workload sample is an important key to determining how well a proposed computer will perform. To develop an accurate workload sample,

top-grade personnel must be given the time and the support to critically review existing and planned workloads on an in-depth basis.

Development of Equipment Specifications

While it may appear to be placing the cart before the horse, it is important that a detailed statement of minimum equipment needs
be developed by the user. Clearly, the user cannot know all of the
products and services that are available from the various vendors
before receiving their proposals. Nonetheless, the user does understand existing and projected workloads, and from this, he or she can
list what the needs will be.

Equipment specifications should be provided in detail (for
example, a minimum-speed-1000-cards-per-minute card reader is
required). It is also a good idea to enclose questionnaires to which
the vendor must respond detailing the equipment characteristics of
the devices, packages, and services being proposed. By using this
approach, the user will be assured that all vendors respond to the
same questions, in the same order, and in the same format. This
simplifies the subsequent comparative evaluation of vendor proposals. Naturally, the vendor should be encouraged to provide any additional unsolicited information which might have a direct bearing
on the final evaluation. For example, the vendor should be encouraged to also demonstrate alternative equipment configurations
which hold promise for improved performance.

Request for Proposal (RFP)

An invitation is sent out to prospective vendors notifying them
simultaneously of an interest in obtaining an equipment proposal
which will meet required specifications. It is important that all vendors are notified at the same time and that they are given a reasonable time within which to respond. While the initial vendor response is simply an indication that they will or will not submit a
proposal, it is important to clearly state that responses received
after the established deadline will not be considered. This approach
sets an atmosphere of fairness and objectivity and also clearly indicates that the vendor must react to *your* timetable throughout the
evaluation process.

The individual(s) to whom all vendor correspondence and proposals must be directed should also be stipulated along with an overall
timetable of events. It is particularly important to notify vendors
what the performance evaluation will be based upon and when
sample workloads may be obtained. The vendor will, in most cases,

need time to convert the representative user workloads to operate on the new computer system.

Proposal Review and Evaluation

As vendor proposals are received, the user must begin immediately to review them in depth. These documents will answer many important questions. Also, equipment features which the user will want to observe "live" will become apparent as a result of this review. Again, top-level computer-user personnel must be given time and support to do an adequate in-depth analysis of these proposals. To wait until the last moment for this review will only compound and delay the evaluation procedure.

Equipment Demonstration and Evaluation

At a mutually agreeable time and location, the user will witness a live demonstration (demo) of the proposed equipment's ability to perform on the representative workload. It is important for the user to structure the approach to this "performance." Often the user plans a team approach, not unlike that employed by the vendors themselves. Each vendor will have a lot of personnel at its demonstration. One or two user representatives cannot hope to get a clear picture of what is happening, or not happening. User personnel should be assigned to monitor at least the following areas:

1. Computer console operation
2. Hardware verification
3. Job timing verification
4. Software verification
5. Peripheral/mainframe interactions
6. Input/output materials collection and control

It is also wise to have a user representative available to discuss other factors such as educational and training materials and management of the proposed computer system.

If we can believe even a fraction of the "war" stories about what can happen at a computer demonstration, then it is a wise user who approaches each demonstration with caution and a sound plan of attack.

Final Evaluation and Selection Proposal

In the final recommendation to management, the user must prepare a complete and objective statement which spells out clearly the reasons for these choices. The factors evaluated and the relative importance of each must be specified along with a comparative analysis of overall vendor performance.

In the section of this chapter entitled "Factors to be Evaluated," the scope of the final proposal will become apparent. The use of a point-scale evaluation technique like the one discussed in that section is highly recommended.

When writing the actual proposal it is a good idea to place a summary of the evaluation and selection process, including the recommended selection, at the front of the proposal. The detail which supports this recommendation should follow and be grouped by evaluation category. Within each category vendor information should appear in order of excellence.

This proposal, once approved by management, will become the basis for notification of all vendors. It is not, however, distributed to vendors, nor should they be allowed to review the document—to do so would be a breach of the confidentiality which each vendor has entrusted to you.

Vendor Notification and Debriefing

Each vendor, win or lose, has invested considerable time, energy, and money in trying to win your business. It is only fitting that they be notified of the outcome quickly and cordially. Provisions should be made for an individual debriefing session with each vendor that did not receive the contract. However, some care should be taken in approaching these debriefings.

First, indicate clearly that the evaluation and selection is complete and that the judgement made on the proposals is final. To discuss "what if's" or last minute "sweeteners" helps no one and can seriously damage your future credibility.

Second, it is important to avoid talking about other people's performances. The discussion should be limited to the advantages and deficiencies of this vendor's proposal as it relates to your needs. If this can't be done, then perhaps the evaluation was not adequately performed.

While the present is unpleasant for the vendors who lost, the future can be consoling. Without overdoing it, the user can encourage them to keep the company abreast of new developments which might place their products in the best position to meet future needs.

Contract Negotiation

The final step in the evaluation and selection procedure will be the negotiation of a contract between the user and the vendor whose equipment is selected. The contract agreement must be both comprehensive and detailed, as it becomes the basis of future dealings between vendor and user. It is, in effect, a mutual commitment and understanding of each party's duties and responsibilities.

From the user's point of view it is advisable to have, clearly stipulated in the agreement, favorable statements regarding:

1. Price protection—Under what conditions can price changes occur and what amount of advance notice is required of the vendor? What discounts apply, if any, and for how long? What will be the pricing arrangements for new or add-on equipment?

2. Overtime usage—What basis, if any, will be used to determine overtime payment by the user? What is the rate for overtime payment?

3. Equipment maintenance—What is the schedule for preventative maintenance of the equipment? What are the conditions which apply to repair service (for example, response time of the vendor to a "trouble" call)? What is the cost of maintenance?

4. System inoperability—Under what conditions will the system be considered inoperable? What payment adjustments, if any, can be enacted as a result of total or partial systems inoperability?

5. Vendor support services—What special services, such as systems and programming support and training, can be expected from the vendor and at what charge to the user? What payment adjustments can be enacted in the event that these services are not forthcoming?

6. Contract termination—Under what conditions may the contract be terminated? How much advanced notice is required by the user to terminate the agreement? What special costs such as shipping charges, must be assumed by the user in the event of contract termination?

The above list includes only a few of the more important items which should be stipulated in the contract agreement. Legal advice from persons familiar with this type of contractual agreement must be obtained by the user prior to signing any written agreement with the vendor.

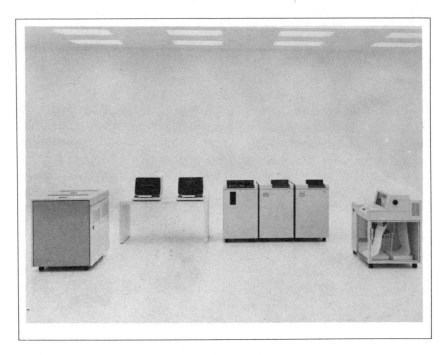

IBM System 8100, a small-scale general-purpose computer

FACTORS TO BE EVALUATED

In addition to equipment performance, several other important factors should be included in the overall evaluation. Computer vendors vary considerably in what they are able or willing to provide to prospective customers. Competition, or the lack of it, in the marketplace may also be a factor in what "can" or "cannot" be obtained from a particular vendor.

Factors of particular interest to most customers include the following:

1. Price/performance factor—An estimate of equipment performance as it relates to the actual cost of the equipment.

2. Systems capability—An evaluation of the hardware and software features of the equipment which have value or potential value to the user and which are not or cannot be fully evaluated according to performance criteria.

3. Systems support—An assessment of the hardware (field engineering) support; systems, programming, and software (sys-

IBM 370/138, a medium-scale commercial-oriented computer

Burroughs B6800, a large-scale computer

tems engineering) support; and training and educational support services provided by the vendor as a part of the standard equipment cost.

Thus, equipment evaluation means the comprehensive evaluation of a complete "package," not simply a piece of equipment. Let us now look at each of these important factors in greater depth.

PRICE/PERFORMANCE FACTOR

Because price is an obvious limit and because it is increasingly important to obtain the "most" (performance) for the "least" (dollars), it is now common practice to arrive at a single value to represent both price and performance. Indeed, many users give both price and performance equal weight in their evaluations. Therefore the computer vendor who offers less capacity at a substantially lower cost can compete with another vendor who offers a little more capacity at a substantially higher cost. While both price and performance are combined into a single price/performance factor for final evaluation, each of these is initially computed separately.

Computer Pricing Arrangements

Computer pricing arrangements have in recent years become quite varied and complex. Not all of the arrangements suggested in this section (see table 10-1) are available from all computer vendors, nor are they available to all customers.

Computer rental agreements are generally arranged so that, for a flat monthly rate, both the use and the maintenance of equipment are covered. In the past, many rental agreements applied to the use of the equipment for a single eight to ten hour shift only. Overtime use of the equipment was either billed at a per hour rate or covered in a supplemental contract for unlimited overtime use. Today, unlimited overtime is often built into the basic per month rental cost. Also, when it is chargeable, overtime does not begin simply at the end of a designated eight-hour shift, but is based on actual systems use during a given month. Under many computer rental plans, a percentage of monthly payments can be applied to the eventual purchase of equipment. This is generally done with some uppermost ceiling (say, up to 50 percent of the purchase price).

Computer purchase plans normally involve a single lump-sum payment for the equipment. Because the user buys the equipment outright, there are no restrictions on amount of use and no monthly

TABLE 10-1   Computer pricing arrangements

Pricing Method	Separate Maintenance Contract Required	Unlimited Overtime Use	Interest Payment	Type of Payment	Discounts	Cancellation Clause
Rental	no	optional	no	monthly	possible	30 - 90 days
Purchase	yes	yes	no	lump sum	usually	none
Lease Purchase	yes	yes	yes	monthly	possible	30 - 90 days

payments for interest or outstanding principal. There is, however, a need to contract for maintenance services. Considerable care must be taken in stipulating the responsibilities of the firm providing these maintenance services (generally, but not necessarily, the vendor), and some recourse for the user in the event of contract default.

In recent years, computer manufacturers have proposed lease-purchase plans to prospective customers. In effect, under the lease-purchase plan the user buys the equipment, but pays for it on a monthly installment basis and with some option for cancellation. A separate maintenance contract must be negotiated, but because the user retains the option to cancel the equipment contract with proper notice, many of the potential difficulties inherent in a purchase plan are removed. Overall, the lease-purchase plan offers some decided advantages to both user and manufacturer. Because installment payments can be extended for as long as seven years, the customer frequently pays less per month (principal and interest) for equipment and maintenance than would be required for normal rental of the same equipment.

As mentioned earlier, the actual cost under any of these plans can vary considerably from one customer to another and from one computer supplier to another. This is primarily due to discounts. Most educational and governmental institutions are able to obtain discounts of some type under any one of the above plans. Many other users can obtain discounts when equipment is purchased or lease purchased.

Another alternative is third-party leasing. Sometimes a company purchases new or used equipment from a manufacturer or other source, taking advantage of considerable discounts or price reductions, solely for the purpose of leasing that equipment to other data processing users. These companies are generally able to pass on some portion of their savings to the customer. Furthermore, in some instances, third-party leasing companies have been able to negotiate with the original manufacturer to maintain the equipment at normal rates once it is installed by the user. For the user who is willing to sacrifice the blanket of security (real or imaginary) which comes with obtaining equipment directly from a manufacturer for the lower price which is available from a third-party leasing company, this is a reasonable alternative.

Obtaining the Best Price

With the many different pricing alternatives available, it is important to ask, "How can I obtain the best possible pricing arrangement for my needs?" In recent years the practice of approaching computer evaluations through "competitive bidding" has gained many adherents. In the public sectors of our economy, such as federal, state, and local government, and in public education, this practice is generally mandated by law. It has also gained many enthusiastic, cost-conscious supporters in the private economic sectors. There are several myths about opening computer procurement for competitive bids which sometimes obscure its true value.

For one thing, competitive bidding does not mean that the user will select the "cheapest" offer. As we shall see, price is only one aspect (and not the most influential) of the total evaluation process.

Second, there is the prevalent notion that better products and services necessarily cost more. Many consumer studies indicate that there are factors other than quality which enable vendors to get more or less money for their products. Computers are no exception.

Lastly, there is the notion that the price on the tag reflects a kind of absolute minimum required by the vendor to make the sale worthwhile. Marketing is far more complex than adding costs, tacking on a fixed profit, and arriving at an absolute price. Marketing considerations are interconnected with such intangibles as consumer attitudes and beliefs, corporate image, and existing competition in the marketplace.

In effect, by opening equipment procurement to competitive bidding, we are inviting computer suppliers to give us what we require at the "best possible" price. We are also advising them ahead of time that they will be judged accordingly.

Methods of Performance Evaluation

The second aspect of the price/performance factor is performance itself. In the past, performance was judged either mathematically or by allowing the computer suppliers to demonstrate equipment with computer programs which they had prepared. In recent years, this approach has been questioned. Vendor-prepared programs rarely, if ever, reflect the user's real workload requirements. Furthermore, equipment suppliers tend to demonstrate things which their equipment does well—not to contrive situations in which the performance of their equipment would be poor or even average. Thus, the burden for developing reliable objective means of performance evaluation has shifted to the customer in recent years. Demonstrating performance capability according to these standards necessarily rests with the computer supplier. There are three commonly used methods of judging performance, namely:

1. Benchmark programs
2. Workload models
3. Computerized workload simulation

Benchmarks   Benchmarks are computer programs representing the user's primary workloads. They may be programs in actual production, or they may have been developed especially to reflect current and proposed processing needs. For example, if a user has a reasonably large number (300–500) of production programs with a high central processor requirement, a program with a very high processor requirement may be selected or constructed as a benchmark. By comparing actual computer performance of the benchmark on the installed system with that of each proposed system, the user is able to obtain an idea of relative performance advantages of each proposed system for this kind of workload.

Workload Models   Workload models are a kind of extension of the benchmark approach. A model is one program or a series of programs weighted to reflect the entire current and projected workload. For example, if it is determined that 20 percent of a user's workload is program compilation, 10 percent is high processor demand workload, 15 percent is sorting, and the remainder is essentially input/output, an appropriate program can be selected or developed to reflect the relative importance of performance in each of these areas. The previous example is an oversimplified, but essentially accurate, view of the concept of modeling.

Workload Simulation Computer workload simulation is the most recently introduced of these three performance assessment methods. Programming packages which perform workload simulation have been developed by a number of companies and have proved to be helpful to a number of users in the evaluation and selection process. One of the more successfully marketed simulation packages was developed by Comress, Inc., of Washington, D.C. It is known as the Systems and Computers Evaluation and Review Technique or SCERT. Initially, the package consisted of four major components.

- The definition language—defines the application to be processed

- The factor library—defines the characteristics of both hardware and software for the computer(s) to be evaluated

- The simulation programs—the computer programs that actually do the simulation

- Output reports— present the results of the evaluation

With the total SCERT package (which can be run on any one of a number of computers), it should be possible to simulate the performance of carefully specified workloads. Critics of SCERT and similar simulation packages generally focus on difficulties with the factor library and the accuracy with which hardware/software characteristics can be translated into performance. Nonetheless, this approach has been heralded by many as the direction which equipment evaluation will increasingly take in the future.

SYSTEMS CAPABILITY FACTOR

In the past, systems capability and performance were frequently confused. With the development of more objective standards for evaluation of performance, systems capability has become visible as a separate important factor to be evaluated. Its two component subfactors are software capability and hardware capability. The evaluation of systems capability is, as we shall see, the attempt to take into consideration features of the proposed system which are not or cannot be fully evaluated according to performance criteria. It is important to verify a manufacturer's statement that these systems capabilities exist, but this can usually be done quite satisfactorily outside of the performance evaluation.

Software Capability

In evaluating software capability we are really attempting to assess at least four software aspects, including:

1. The operating system
2. The available programming languages
3. The utility programs supplied by the vendor
4. The application packages supplied by the vendor

Any exhaustive explanation of the evaluation process for each of these software features clearly exceeds the intentions of this single chapter on evaluation and selection. However, some brief comments are possible and, indeed, important within the set limits of time and space.

An operating system is a master or supervisory program, usually provided by the vendor, which enables the computer system to operate as a system. Operating systems vary greatly in comprehensiveness and operational efficiency. The primary question is: How well does the operating system interact so as to ease the burdens of operating, programming, and managing the computer system? Another is: To what extent does the operating system lead to efficient utilization of existing capabilities and technological advancements? For example, attention should be given to such vital areas in the operating system as:

- *Job scheduling*   How effectively does the machine work to optimize internal utilization of its resources?
- *Disk and main memory utilization*   How are these resources allocated and maintained?
- *Processor utilization*   How effectively is the C.P.U. utilized?
- *Operational aspects*   Does the operating system permit relative ease in operating the computer?
- *Programmatic aspects*   How readily does the operating system accommodate requests for programming options, program segmentation, and overlaying?

From the limited list above, it should be apparent that there are many operating systems capabilities which may make a significant difference when a computer is used on a day-to-day basis or for an extended period of time.

Available programming languages are evaluated from a number of vantage points. It is important to know what kind of compilers are being proposed and whether or not they meet minimum national standards (if they exist) or local requirements. For example, is ANS COBOL available? Frequently, subset or limited instruction versions of various languages are also available, which may or may not be advantageous depending on user need.

In addition, it is important to determine compiler minimum main-memory requirements and other compiler characteristics. Compiler diagnostics and documentation are two areas of extreme importance which will affect experienced programmer productivity and the training of new programmers.

Utility programs are generalized programs which perform basic data processing functions, such as card-to-print, tape-to-tape, and tape-to-print. They cover the full spectrum of possibilities from programs which perform system diagnostics to those that simply load and unload data. Evaluation of these utilities should ascertain the following:

• For what specific functions are there utility programs?
• How efficiently do they operate?
• How easy are they to execute?
• How flexible are they in accepting a wide variety of input/output formats?

A sufficient variety of well-conceived utility programs will save valuable programmer, operator, and machine time.

An application package is a series of vendor-prepared computer programs designed to fulfill major needs for some users. A series of vendor-written programs which could be used in a hospital administration system is an example. Application packages represent one of the biggest "if's" in the computer evaluation process. Certainly, the amount of value given to available application packages will vary greatly, depending upon current or projected user needs. For a user who is in the hospital business, a superb package of hospital administration application programs available as part of equipment rental may be a significant evaluation factor in and of itself. Naturally, if the user is not in the hospital business, that same package loses significance.

A few application packages have some relevance for most if not all computer users. One is a true job-logging or job-accounting system. This package will enable the computer to maintain records of its own utilization by creating a "log" of all jobs run. The package will also contain the necessary programs for preparing customer bills for computer services.

Others are file and data-base management packages. These packages can save many hours of top-level programmer time for the user. Smaller computer installations frequently cannot afford specialists who could develop, on their own, similar programs. For them file and data-base management application packages are even more important.

Hardware Capability

Systems hardware capabilities can be evaluated more directly than those of systems software. Many of the hardware requirements will have been detailed in the set of equipment specifications prepared by the user and given to the vendor as part of the request for proposals. In general, the user will be asking:

- Did the vendor meet the minimum hardware specifications?
- Is the system truly expandable in a way which meets longer-range needs?
- Will hardware expansion mean faster throughput or simply greater capacity?
- Did the vendor exceed equipment specifications to an extent worthy of special consideration?

SYSTEMS SUPPORT FACTOR

In the evaluation of a new systems proposal, users generally focus on three primary aspects of systems support: field engineering (hardware), systems engineering (systems and programming), and education and training programs.

Field engineering support may be the most consistently undervalued service offered by a vendor. In the excitement of the computer demonstration and sales presentations, the engineers who set up, repair, and maintain the equipment are frequently overlooked. Once the marketing and sales personnel have left in their quest for bigger and more profitable accounts, the quality of the field engineering support becomes important.

To properly evaluate the proposed field engineering support, the following areas can be crucial:

- *Preventative maintenance schedule* How frequently and to what extent will your equipment receive maintenance attention before breakdown?
- *Assignment of field engineering personnel* Who are they? Where do they travel from? How many other accounts, if any, are they committed to?
- *Parts inventory* What arrangement will be made to store crucial equipment parts on site or at a nearby location?
- *Response time to user calls* What is the maximum field engineering response time to be expected if and when the system goes down?

- *Overtime charges* What are the normal maintenance times? Under what conditions may the user be charged overtime?
- *Specialist support levels* Under what conditions will higher-level engineering personnel be consulted or brought on site to deal with extraordinary hardware difficulties?
- *Back-up computers* Under what conditions can free back-up computer time be made available to the user and what are the alternative locations?
- *System downtime/system payment correlation* What are the minimal performance levels for which full payment must be given to the vendor? When performance falls below these levels, what will be the percentage of payment, if any?

The use of free systems engineering services has traditionally been a valuable asset to the computer user, particularly because new computers frequently imply large-scale one-time systems and programming-conversion efforts. These efforts can easily exceed the capabilities of the user's regular staff. The temporary assignment of professional systems and programming personnel employed by the vendor can be very valuable, particularly when the user department has a small to medium-scale computer installation.

Beyond this temporary initial hurdle, the user should be concerned about continuing access to vendor systems and programming personnel for consultation when and if software modifications and enhancements are introduced. A few hours of consultation with a knowledgeable systems engineer may save days, weeks, or even months of precious staff time.

For evaluation, it is important to know who will provide the support and the extent of the support commitment. Many evaluators insist on receiving detailed work resumes for the individuals being assigned as systems support staff by the vendor. If systems support staff are to be assigned on a full-time basis (on site) during the early months, this must be clearly stipulated in the negotiated contract. It is also a good idea to clarify the meaning of "continuing support," which can be anything from a limited continuation of on-site support (for example, two person-days per month) to someone available at the other end of a telephone whenever problems arise.

On occasion the level of a technical problem may exceed the knowledge and understanding of the systems engineering personnel assigned to an account. At times such as this it is important to have access to "home office" experts, particularly with software difficulties. A procedure which spells out the kind of situations in which back-up high-level technical expertise can be made avail-

Field engineering service repairing CPU

able to the user represents a sound approach to dealing with this important area.

Training and education can be broken into three important areas: formal courses, vendor-training materials, and documentation. The formal-course commitment varies widely from vendor to vendor and from one account to another. At one end of the spectrum, courses for which the user must pay are offered at particular training centers, say in Boston, Los Angeles, and Chicago. At the other extreme, some vendors offer "no pay" courses on site, providing the user can meet a minimum enrollment of eight to ten personnel. Most likely, there is a combination of on-site and training-center offerings, some of which are without charge to the user.

Training materials in the form of books, video tapes, film strips, and so on have become an increasingly important means of user training. While these materials may be costly, the highly professional nature of these presentations merits close scrutiny by the user. To some extent the equipment which a user selects is important in this respect. If the proposed equipment is widely installed, the possibility of finding vendor or independently prepared training materials is considerably greater than if the proposed equipment is either highly specialized or used by only a few customers.

Complete up-to-date documentation on the systems hardware, software, and application packages is vital to a user's ultimate success or failure in using a computer. This is an area in which many vendors have traditionally fallen short of user expectations, not just in the quantity of available documentation, but also in the quality. One very good reason for including documentation in the overall computer evaluation is to inform vendors that this is an area of keen user interest. Perhaps in this way some added degree of vendor emphasis on quality documentation can be stimulated.

The support services offered as a regular part of equipment procurement (at no extra charge) vary widely from vendor to vendor and perhaps even more so from contract to contract. Many vendors today use separate pricing, more commonly referred to as *unbundling*. With unbundling, many support services are not available without additional charge beyond basic rental costs. Although vendors did, in the process of unbundling, lower rental prices by several percent, critics argue that (1) the price cuts were temporary, and (2) the loss of services was far more costly to the user than the downward price adjustments reflected.

A second factor accounting for variation in support services is availability of vendor support staff. Many vendor support-group leaders complain that marketing consistently outpaces their ability to hire and train needed personnel.

Local factors also enter into the problem. Many of the smaller vendors find it difficult to propose extensive support services in remote areas and in areas where they do not have the customer base needed to make extended support economically feasible.

MAKING THE SELECTION

Hopefully the preceding discussions have encouraged the reader to think more carefully about the equipment evaluation process. There are any number of ways in which the various factors and subfactors discussed may be combined to produce a single gauge upon which the final selection is based. One way in which these factors might be combined is presented below. Particular user needs and

Factor	Maximum Numerical Score
Price/Performance	35
Systems Capability	50
Support	15
Total	100

Figure 10-1   Relative weights for three major factors on a 100-point scale

circumstances could require some variation from these suggested guidelines.

On a 100-point scale, the primary factors to be weighted are: (1) price/performance, (2) systems capability, and (3) support. Note (see figure 10-1) that while the price/performance factor is large enough to easily affect the overall evaluation, it is not the largest single factor. Consider for a moment that the performance criteria are, in fact, a sample of what a user does and wants to do based on current estimates and past experience. It would be unwise to presume 100 percent predictability based on these important but limiting criteria. By and large, it will be the availability of important and innovative hardware and software capabilities which will offer significant new design and operating alternatives.

It is in no way demeaning to assign a relatively lower level of importance to the support factor. A quick review of the relative weights suggested in figure 10-2 further indicates that the primary user emphasis should be in the field engineering support area. Overall, it is likely that a poor performance in this area will drastically reduce a vendor's chances of coming out on top in the evaluation.

Factor/Subfactor	Subfactor Maximum	Factor Maximum
Support		15
Field Engineering (FE)	50	
Systems Engineering (SE)	15	
Education and Training	35	
	100	

Figure 10-2   Relative weights for support subfactors

Factor/Subfactor/Category	Category Maximum	Subfactor Maximum	Factor Maximum
Systems Capability			50
Software		55	
Operating System	45		
Languages	25		
Utilities	10		
Application Packages	20		
	100		
Hardware		45	
		100	

Figure 10-3   Relative weights for systems capabilities subfactors and categories

A further breakdown of the systems capability factor (see figure 10-3) suggests a 55–45 respective weighting of software and hardware. In recent years, the importance of effective software has caused many evaluators to assign it a slightly higher weight than hardware.

Factor/Subfactor/Category	Category Maximum	Subfactor Maximum	Factor Maximum
Price/Performance			35
Performance on benchmarks		50	
Price on configuration		50	
		100	
Systems Capability			50
Software		55	
Operating System	45		
Languages	25		
Utilities	10		
Application Packages	20		
	100		
Hardware		45	
		100	
Support			15
FE Support		50	
SE Support		15	
Education and Training		35	
		100	100

Figure 10-4   Suggested numerical scoring system

By using the suggested weighting scheme it becomes practicable to:

1. Notify the competing vendors of how they can expect to be evaluated before they begin.

2. Assign a numerical score for each factor.

3. Tell how each vendor performed for each variable.

4. Determine with relative ease a single score which is based on comprehensive multiple variables.

The completed numerical scoring scheme is probably best explained graphically (see figure 10-4).

MINICOMPUTERS AND MICROCOMPUTERS

The most difficult questions to answer about computer hardware are frequently the most basic ones, such as, "What is the difference between a minicomputer and a small, general-purpose one?" or "What is the difference between a minicomputer and a microcomputer?" In fact, the lines of technological difference between these kinds of computers have blurred to the point where one feels, at times, that the true difference is one of different marketing strategies among competing vendors.

Admitting that there are no all-inclusive answers to the above questions, we may cast some light on the issue by looking at the historical emergence of minicomputers and microcomputers and what they sought to achieve that was different.

The minicomputer, as its name implies, was introduced as a computer that was to be smaller than others in several ways, namely:

- It cost less.
- It was physically smaller.
- It was frequently designed to be application-specific (or was smaller in the sense of proposing to do fewer kinds of data processing than its general-purpose counterparts).
- It was frequently equipped with less powerful processing and fewer peripheral capabilities (but these capabilities were sufficient to meet the needs of many prospective computer users).
- It was also "smaller" in the resources required to support it, such as air conditioning, computer operations staff, maintenance personnel, and so on.

Today, so-called "minicomputers" range in size, span of function, processing power, and support requirements. We could say that the true minicomputer is one that scores at the low end for each of the above measures, were it not for the more recent introduction of microcomputers.

Microcomputers, or micros, evolved as a form of desktop computer with only limited memory and peripheral capacity, and frequently with the entire CPU on a single memory "chip." The micro's introduction marked an important stride forward in the development of the personal computer. The typical micro initially looked like a CRT terminal (or a keyboard that could be attached to your home television) with an added "black box" or two that were, in fact, flexible or "floppy" disk drives. Typically, these early micros permitted programming in interactive languages like BASIC and lacked the capacity to perform COBOL or full FORTRAN compilations. Consequently, the early micro was more suitable as a device for learning about computing than as a true business data processing machine.

Again, time and technology have begun to change this situation. Today, an increasingly broad range of power is available in the micro marketplace. An increasingly broad range of peripherals and applications is also available.

The previous sections of this chapter presented a rather comprehensive outline of how to evaluate and select computer systems. Although that approach (and the time and effort required to execute it) would be appropriate for a $5 million expenditure, on the surface its value would appear questionable for acquiring a $500 device.

Let us look at some of the pitfalls frequently encountered in acquiring minicomputers, or minis. First, although the minicomputer may have been designed around a particular function that it performs very efficiently, it can generally perform or be made to perform other functions (although very inefficiently). Therefore, it is particularly important that a complete analysis and performance evaluation be conducted for all workloads.

Second, minicomputers are frequently acquired by organizations with little or no in-house computer experience. In such cases, evaluation factors such as ease of use, vendor-maintained software, user training, and other nonperformance factors should be weighted very highly indeed. Finally, we know that minis come in a variety of packages, each differing in options, performance level, and price tag.

To avoid the above pitfalls, and particularly to aid the organization that lacks an in-house data processing function, the formal evaluation and selection process is recommended. When the equipment being sought is at the lower end of the spectrum, the process can be less extensive, but the implications of each step in the process must still be reviewed.

Microcomputer evaluation and selection are performed in the formal sense only when a large number of micros are being acquired simultaneously. When one or a few micros are being acquired, a much less formal process is frequently used. Nonetheless, several points of caution need to be emphasized for persons acquiring a microcomputer:

- Critically evaluate sales promises in much the same way that was suggested for formal evaluation.

- Business application software for microcomputers has usually been developed independently of the computer manufacturer. Be sure that such software is available and meets your needs in its current form.

- Maintenance and needed repairs may be obtainable from your dealer or (by returning the device) from the manufacturer. Understand what your service options are and plan for those times when the device is being repaired.

- Low-end, minimally configured micros are priced impressively low, but the added components you will require to perform a business function may double or even triple the price.

- Prices vary widely for the same level of microcomputer capability. It is worth conducting at least a modest exploration of what is available.

One step in formal computer evaluation and selection should always be exercised, namely, the live demonstration of the proposed hardware and software configuration. If the proposed configuration can accommodate a payroll for twenty-five employees, then clearly demonstrating this before acquisition should be no extensive exercise.

CHAPTER SUMMARY

The evaluation and selection of computing equipment is best approached as a formal process with the following steps: (1) workload analysis, (2) development of equipment specifications, (3) request for proposal (RFP), (4) proposal review and evaluation, (5) equipment demonstration and evaluation, (6) final evaluation and selection, (7) vendor notification and debriefing, and (8) contract negotiation.

The final evaluation and selection should give major attention to three primary factors: price/performance, systems capability, and systems support.

To obtain the best possible price for equipment, many users approach equipment procurement through a competitive bidding process. By considering both price and performance as a single factor and practicing competitive bidding, the user eliminates the possibility that equipment will be selected simply because it is cheaper or because it outperforms all competitors (although at a higher cost).

Demonstrations of equipment performance are generally approached using one of the following methods: (1) benchmark programs; (2) workload models; or (3) workload simulations.

Systems capabilities are usually evaluated as a separate factor since they are frequently difficult to test as a part of equipment performance. Software capability is a function of the operating system, programming languages, utility programs, and application packages, which operate on the equipment. Hardware capabilities include many important aspects of the equipment, such as capacity, expandability, and hardware limitations.

Systems support is evaluated on the basis of the kind and extent of field engineering (hardware), systems engineering (systems and programming) and training and education support services that are provided by the vendor.

A point-scale approach to evaluation and selection of equipment is recommended. Such an approach provides an important basis for comparative evaluation and results in a single final score which is based on multiple variables.

The advent of minicomputers and microcomputers does not obviate the need for thorough evaluation to achieve an informed decision. The basic factors to be explored remain constant, although their relative weightings in the evaluation may vary.

WORD LIST

benchmarks—computer programs that represent the user's primary workloads and are used in the performance evaluation of computing equipment

job accounting system—a series of computer programs which enable a computer to maintain records of and report on its own utilization

price/performance factor—a single-value expression of equipment performance as it relates to the actual cost of the equipment

request for proposal (RFP)—an invitation from the user to computer vendors to submit an equipment proposal

SCERT—Systems and Computers Evaluation and Review Technique, which is a computer workload simulation technique developed by Conress, Inc.

system downgrading—modification or replacement of the existing central processing unit (CPU) or computer mainframe with a resulting decrease in computing capacity

system upgrading—modification or replacement of the existing central processing unit (CPU) or computer mainframe with a resulting increase in computing capacity

workload model—a program or series of programs used in computer performance evaluation which is weighted to represent the entire current and projected user workload

QUESTIONS AND PROBLEMS

1. Problem:
    Pseudo Instruments, Inc. would like to increase its computing capability. Harlan Simpson, the Business and Finance Manager, has taken the position that vendor Y should be approached since it is the largest (and to his thinking the most prestigious) computer supplier in the country. Sally Pierson, Assistant to the President, has pointed out that a major competitor has experienced numerous problems with Y's new line of computers. She feels that Vendor X which has been getting a lot of recent praise in the trade magazines should be approached first.

    The Computer Center Manager, Angelo Cocca, has pointed out that a change in equipment vendors (currently vendor Z's equipment is installed) will place a great burden on the company's small computer staff and will result in significant additional cost to company.

    As a consulting systems analyst how do you respond to each of the concerns expressed above? How do you recommend that Pseudo Instruments approach the problem and why?

2. After extensive evaluation, you have determined that Excell Computer's model V-29 is best for your firm. During the vendor debriefing, a competitor schedules a meeting with your senior vice-president and challenges your technical competence.

    *   What might you have done in anticipation of this action?
    *   How should you deal with the action, now that it has unexpectedly occurred?

3. The Research and Development Unit in your company has decided to acquire a minicomputer that is especially well suited to research scientists' needs. You are asked to brief the R&D managers on how to approach the selection.

   Develop a topical outline and appropriate illustrations for your presentation.

4. Three divisions in Xon Corporation have installed Zero 9754 computers. These machines are no longer marketed by Zero Corporation, and although parts have been difficult to obtain in recent years, the machines have always performed adequately. Dramatic increases in the San Francisco division's sales volume have caused that division to acquire a more powerful computer and offer its 9754 to the other two divisions.

   The Chicago and New York divisions make a strong case to management underscoring the critical shortage of parts.

   You are asked for an opinion on what the firm should do. What concerns would you have? What would you recommend, and why?

5. Briefly explain how each of the following relates to equipment evaluation and selection:

   a. Price performance factor
   b. Systems capability factor
   c. Systems support factor

6. You are a systems consultant to a small manufacturing firm located in northern Maine. How might the firm's location and size influence your weighting of the various evaluation and selection factors?

MINICASE—CONTEXT INTERNATIONAL, LTD.

The new financial system will apparently require a combination of enhanced and new computing hardware and software. There will probably be considerable disagreement among certain subsidiaries' data processing personnel about what equipment to acquire.

   You are aware of—and particularly sensitive to—the fact that one subsidiary strongly favors continuing to use CTU computers, and another equally strongly favors the new TNT model 50.

Recognizing that a unilateral decision by you would be least acceptable, you must structure the evaluation to obtain participation from the subsidiaries' data processing personnel as well as top management.

In a team approach, who should be on each team? (Refer to the minicase at the conclusion of previous chapters for additional background information.)

1. How might you organize the work around several teams?
2. How would you control the process?
3. What would be the potential advantages and disadvantages of using such an approach?

11

Project Management
and Control

CHAPTER OBJECTIVES

When you finish this chapter you will be able to—

- *Discuss the process of task analysis including the steps, the method, and the purpose.*
- *Discuss the process of estimating time schedules, which includes developing job and personnel standards and assessing people and jobs.*
- *Discuss the requirements and techniques for effective project control.*
- *Compare and contrast Gantt Charts versus the Critical Path Method (CPM) as project management and control tools.*
- *Diagram a CPM network, given an activity table.*
- *Compute the EST, EFT, LST, LFT, and the float (slack) time for each activity and determine the critical path for a given problem.*
- *Discuss the extensions of CPM typically associated with the Program Evaluation and Review Technique (PERT).*

A systems analyst may have the opportunity to act as a project leader or project manager. For those who enjoy the application of management techniques to systems analysis, this will be an opportunity for both personal challenge and reward.

A project manager assumes overall responsibility for the completion of a systems project from problem definition to implementation of the final system. In most instances, the project manager will have to supervise the activities of other systems analysts, ensure that work is completed according to schedule, and coordinate the activities of the project team with the needs and demands of the user for whom the system is being developed.

To ensure that the project team will effectively meet the needs and demands of the user, the project manager must be skilled in:

1. Scheduling and allocating human and equipment resources
2. Controlling or monitoring project progress
3. Initiating corrective action when it is required

These three activities are interdependent. Effective scheduling, is not done once and forgotten. After checking upon actual progress, the project manager must frequently reschedule personnel and equipment. Similarly, it is impossible to adequately control or monitor an unscheduled project, because it has no structure within which to measure actual progress.

To understand how the project manager acquires skill in each of these three vital areas, let us consider some of the major problems confronting the manager and some of the widely accepted methods of dealing with them. Later in the chapter two different systems tools which attempt to facilitate dealing with many of these problems will be discussed, namely, Gantt Charts and the Critical Path Method.

The Order Within Jobs

The project manager must determine not only the logical order in which jobs should be done, but also the steps in each job. It is not enough to say that a particular machine must be installed before work on it can begin if one is to adequately schedule and control its installation. Rather, it is necessary to detail, in logical order, the steps or tasks involved in its installation.

This process of detailing jobs into a series of logically dependent steps is called *task analysis*. Through task analysis the project manager is able to:

- Define the tasks to be done
- Determine the order in which all tasks should be done
- Establish when, if ever, two or more tasks may be done at the same time
- Determine the exact status of a particular job

There is no absolute limit to the extent to which various jobs should be detailed. To a large extent it depends upon the project manager's ability to keep abreast of what is happening at any given level of detail. If computerized assistance is available for project monitoring, then the manager can go into greater task-analysis detail than would otherwise be possible.

Consider a rather simplified example of how task analysis works for the job called "computer programming." Through analysis, one could conclude that it consists of at least the following tasks:

1. Logical analysis (including flow-charting and field definition)

2. Coding (including some debugging)

3. Testing (including some comparative analysis with manual operations)
4. Documentation (including runbooks, and so on)
5. Production test acceptance

A project manager would need to know not only that a programmer was working on a computer program, but also what stage of completion he or she had reached. To further analyze the various tasks involved in computer programming, the following questions should be applied to each task:

1. What precedes it?—Which tasks must be completed before this task can begin?
2. What follows it?—Which tasks cannot start until this task is completed?
3. What can be done concurrently?—Which other tasks can be performed at the same time?

In order to organize the answers to these questions, we can construct a planning table which lists each task and relates it to those tasks which directly precede, follow, or are performed concurrently with it. Figure 11-1 illustrates how our simplified example of a computer program might have been entered into a planning table.

Planning Table

Code Name	Task Description	Task Directly Preceding	Task Directly Following	Concurrent Tasks
A	Logical Analysis	——	Coding	Documentation
B	Coding	Logical Analysis	Testing	Documentation
C	Testing	Coding	Acceptance Test	Documentation
D	Documentation	——	Acceptance Test	Logical Analysis Coding Testing
E	Acceptance Test	Testing Documentation	—	——

Figure 11-1   Task analysis of a programming job

Tasks to be compared	TASK A	TASK B	TASK C	TASK D	TASK E	TASK F	TASK G	TASK H	TASK I	TASK J	TASK K	TASK L
TASK A	•	F										
TASK B	P	•	C									
TASK C		C	•									
TASK D				•								
TASK E					•							
TASK F						•						
TASK G							•					
TASK H								•				
TASK I									•			
TASK J										•		
TASK K											•	
TASK L												•

Figure 11-2   Dependence matrix

Because most projects consist of many more tasks than our previous example and because the relationships between these tasks are sometimes less apparent, it is usually necessary to systematically analyze the relationship of each task to every other task—each time asking if this task directly precedes, directly follows, or is concurrent with the other. To simplify the recording of this information, a dependence matrix such as that shown in figure 11-2 is used before attempting to build a planning table.

In the matrix the name of each task is listed first across the top and then along the left-hand edge. Each task along the top is compared with each one along the left edge (except itself) and a determination is made regarding their relationship. These are the rules for establishing relationships:

1. If the task in the top row must be completed before work can begin on the other task, enter a P for *Precedes* directly onto the grid.

2. If the task in the top row cannot be started until the other task has been completed, enter an F for *Follows* directly onto the grid.

3. If the two tasks may be performed at the same time, enter a C for *Concurrent* onto the grid.

4. If the task at the left-hand edge does not immediately precede or follow and if it is not known to be concurrent, then the matrix grid is left blank or a dot is entered.

The entries in figure 11-2, for example, illustrate that Task A precedes Task B, Task B follows Task A, and Tasks B and C are concurrent.

By using this method, one is able to obtain a true ordering of tasks in sufficient detail for project scheduling and project control.

The Process of Estimation

A good part of the difficulty in scheduling people has to do with estimating how long it will take them to perform a particular task or series of tasks. It is difficult enough to estimate our own performance on a single task. When we add to that the individual differences within a project team and the variety of tasks to be performed, the complexities of the project manager's job become apparent.

While it is true that the process of estimation is an inexact science, it is not entirely "hit or miss." Experience is an important part of making a refined judgement about how long it should take a particular employee to complete a particular task. But experience alone will not yield accurate estimates. The problem for the project manager is to "structure" the situation. Structuring is an attempt to establish a set of standards or guidelines about job difficulty and personnel skill levels in general. Using these standards, actual jobs and people can be evaluated and compared to them. Depending upon the results of this comparison, an estimation of required completion time will be developed. Once actual work begins, feedback will enable us to determine where we went wrong (Are the standards wrong? Did we misjudge the complexity of the specific job or the talents of the assigned personnel?). Figure 11-3 illustrates this general concept.

For the analyst who is assigned as a project manager for the first time, the most difficult aspect of estimation is the development of usable standards.

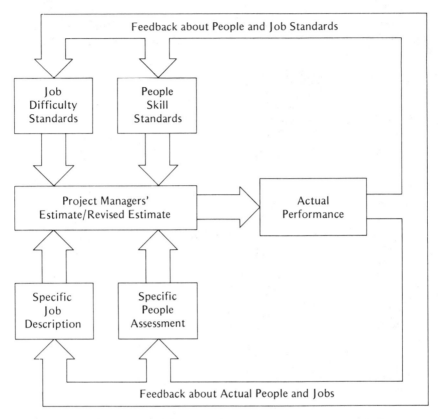

Figure 11-3   Process of estimation

THE DEVELOPMENT OF STANDARDS

In order to better understand how one goes about developing standards, we will use several very general examples. Let us continue with the computer programming example, which has already been broken down into its component tasks.

In order to establish standards for computer programming, it will be necessary to envision the full range of possibilities for:

Job Standards
1. Level of Complexity
2. Level of Problem Definition

Personnel Standards
1. Level of Programmer Skill
2. User Responsiveness/Involvement

Standards for Jobs

The complexity level of a given computer program is a debatable matter, particularly since people confuse level of program complexity with programmer skill. By treating programmer skill level as a second factor, however, one should be able to develop a fairly objective set of standards whereby complexity can be readily assessed. Figure 11-4 illustrates a set of suggested guidelines for computer program complexity. It should be noted that as the level of program difficulty increases, not all programming tasks will necessarily require more time. For example, documentation time is probably about the same for programs, regardless of complexity.

The thoroughness of problem definition is a second major consideration in estimating programming time. Given an extensive definition of the problem, a programmer spends less time on logical analysis. Also, the amount of time spent testing the system may be significantly affected because there will be few, if any, surprises. Standards for a very well-defined problem might include:

Very Well-Defined   The programmer is provided with at least the following:
a) All input and output specifications
b) All file specifications including recommended data names, file size, information source, accessing method, and so on.
c) Detailed logical description of the program functions to be coded and the alternatives which must be accounted for
d) Program test data, which can be used to verify that the program does, in fact, function properly

---

Standards for Estimating Computer Program Complexity
1. Report program with little calculation; for example, tape print out, card to tape, tape to disk
2. Update program, transaction update-sort-report type program
3. Typical program within a system which might include intermediate calculation, extensive table look-up, report presentation
4. Major or key program within a system, which might include difficult calculations and data manipulations
5. Unusual or special type of program, such as a communications data handler or a software modification

---

Figure 11-4   Guidelines for program complexity

To the degree that these materials are not made available to the programmer, the problem is considered to be less well defined, and the amount of time estimated to complete the job must necessarily increase. The "worst" case is probably when the programmer is required to sit down with systems and user personnel to obtain first-hand a definition of the problem. At such times the computer programmer is actually being requested to perform many of the functions of systems analysis. This can greatly expand the amount of time required to do the job and can seriously affect the quality of the final product, since systems analysis is a process for which the programmer has not been adequately trained.

Standards for People

Developing standards to assess people can be dangerous if it is done blindly. For one thing people seldom fit into convenient categories. This is not to say that it is unimportant to develop standards, but rather that it is wise to apply standards with great caution.

It is possible to discuss job standards objectively without regard for the people who will do the work. But personnel standards must be related to the job. Thus to talk about a programmer skill level which is average, the manager must specify what the average skill level is for a particular kind of program (such as a report program with little calculation).

Another consideration has to do with the relationship between the level of skill and the specific task. For example, a programmer may have skill in coding, testing, and documentation, but not in performing logical analysis. Thus the assessment of programmer skill may read as follows:

Programmer has average skill when

- The program to be written is of the simplest kind, (such as, a report program) and
- When the problem definition is very good (there is little need for logical analysis)

In other words, criteria for skill level must be established in terms of both the job to be done and the tasks to be performed. Remembering this, we can develop workable skill standards which may be applied to each kind of task/job situation.

Note that there has been no attempt to correlate overall years of experience, job title, or pay scale with programmer skill. Many project managers make the mistake of equating years of programming experience (as well as the other factors mentioned) with ability to do a specific kind of work. This assumption is a false one and can get a manager into difficulty. For example, if one has been programming report programs for ten years with great skill, that does not mean that one is skilled in writing communications-data-handler programs. A year or two of highly specialized experience in programming communications systems may enable a person with less overall experience, lower salary, and a less prestigious title to be more qualified for that particular job.

A second factor for which personnel standards should be developed has to do with user responsiveness. In addition to the skill of project team members, the project manager must take into consideration the following:

1. How responsive is the user to the implementation of the entire system and to the specific systems component (programming job) in question?

2. How actively involved has the user been in the project in terms of direct participation, people assignment, and other expressions of interest and concern?

3. What kind of people have been assigned to the project by the user? Are they informed? Do they possess authority to make decisions?

Clearly, user responsiveness, involvement, and ability affect the amount of time required by a skilled programmer to complete a job. The amount of time required to go back and re-program because the user has not adequately explained processing requirements can be a significant portion of the total job-completion time.

If the project manager establishes these kinds of standards for people and jobs, then he or she has a frame of reference which will serve as a guide in assessing individual situations. With a structured approach to estimation, experience will serve as the basis for modifying existing standards in a way which will improve the manager's ability to make future estimates.

BEFORE APPLYING STANDARDS

Up to this point the discussion has dealt with establishing standards for job and personnel performance against which real jobs and real people can be compared. Necessarily, the evaluation of the real world, even with carefully established guidelines, is seldom as simple as plugging the right number into the right formula. Before a proper assessment can be made of either people or jobs, a manager must know a good deal about them.

Too often, project managers sacrifice their management responsibilities for other duties and responsibilities, such as saving a "little" systems design or computer programming work for themselves. Project management is generally a full-time job if it is to be done well; these auxiliary assignments can only interfere with a good management job. Nowhere is this more evident than in the assessment of people and tasks.

It is well to recall at this time the prior discussions on effective communications in systems analysis. In particular, people who are working with problems on a day-to-day basis are the primary source of information upon which systems decisions will be made. This also applies to good project management. An effective manager relies on the people who do the work and must, therefore, take the time to listen to their assessments of a problem or a person. It is only after making a personal assessment and giving due consideration to the assessments of others that a management decision should be made.

The manager will gain some interesting insights into the judgements which others make about people and problems. For example, we may find that Programmer A, when asked to assess the time required to complete a programming job, is consistently over-optimistic. Programmer B, on the other hand, may consistently over-estimate the amount of time required to do a job.

It is well to remember that a project manager is, in many ways, primarily a people manager. Therefore, he or she has a responsibility to listen to the other employees and to help them see the different sides of their individual assessments and recommendations.

Confronting Difficulties

Because the systems process is to a large extent characterized by the interaction of people and ideas, we can anticipate many points of disagreement and conflict in developing a new system. It is the project manager who is perhaps best able to determine how to deal with the disagreement.

While many kinds of disagreement and conflict can occur, some are more commonly encountered in systems analysis. These typical problems and some suggestions for dealing with them are given below.

Problem:
Personal disagreement or conflict—either between two participating users or between systems and user personnel

Suggestions:
1. Attempt to determine if the conflict is the result of misunderstanding, in which case it can probably be mediated by the project leader.
2. If the conflict is more a consequence of a personality clash or unmendable differences, attempt to restructure staffing so that work will not depend upon the direct cooperation of the two parties.

Problem:
Territorial disagreement or conflict—for example, the legitimacy of permitting one person to comment upon another's area of responsibility is questioned. A good example is frequently found in systems studies which result in questions concerning administrative reorganization as one aspect of developing new systems alternatives.

Suggestions:
1. View the situation as one in which a person's authority has been challenged or threatened. Anticipate that this kind of conflict could, if left unattended, result in unnecessarily politicizing the systems process.
2. Work with the party who is being challenged to develop some set of acceptable ground rules for investigation of his or her areas of responsibility, including:
   a) What points are not open to question and why not?
   b) What should be the procedure for investigating and commenting upon those areas for which he or she is responsible?
   c) What should be the basis for making decisions affecting his or her areas of responsibility?
3. Plan frequent periodic reviews in these situations and direct them toward confronting the issue. In this way it is less likely that persons will seek to resolve their differences external to the process (through the political manipulation of higher level managers).

Problem:
Ideological disagreement or conflict—for example, when persons differ in their views of what important "issues" should be dealt with. A good example of this might be found in a College Financial Aids System in which one person tends to focus exclusively on the financial accountability and another on the student counselling and assistance side of the problem.

Suggestions:
1. Recognize that there is probably some merit in each of the two opposing viewpoints.
2. View the situation essentially as one which requires negotiation between both parties.
3. Expect that some compromise will probably be required of *both* parties if an acceptable solution is to be reached.
4. Recognize that the ideas have been stated in their most extreme form and for that reason each party can probably compromise without losing what it seeks to maintain.

The primary point to be made at this time is that disagreements and conflict are a natural everyday occurrence in any dynamic organization. The systems process heightens rather than lessens the possibility for disagreement and conflict because it focuses the attention of many people upon specific problems and procedures. The project leader must confront rather than ignore these realities. In working with them the project leader seeks to resolve some differences and to simply direct others toward positive change rather than open warfare.

PROJECT CONTROL

Project control is one aspect of the overall process of project management. Project control is the means whereby the manager can assess:

1. What is happening for any aspect of the project?
2. What relationships exist between work being performed on different aspects of the project?

Good project control begins with the assignment of job responsibilities. It is a good idea to have available some kind of job assignment sheet. This job assignment sheet should contain at least the following:

1. Name and title of the person assigned
2. Date project assigned
3. Name and job number for the job being performed
4. User name and charge code
5. Total estimated completion time for the job (also estimate revisions)

6. Detailed breakdown of the job into its components
7. Estimated completion date for each task (also estimate revisions)

The other major requirement for effective project control is feedback regarding actual job performance. Some project managers like to receive a report (usually called a Project Status Report) on a weekly or biweekly basis from each member of the staff. However, there is no reason why, in most instances, both job assignment and reporting information cannot be contained on the same form (see figure 11-5).

A project manager is primarily accountable for the success or the failure of a project. If there are difficulties in meeting expectations, then the manager must understand the problems in detail. The one or two sentence explanation on a project status report will not be enough. For this reason, it is wise for the manager to discuss problem areas in greater depth with the assigned person.

The third ingredient in effective project control is well-enacted procedures for job assignment and reporting job status. The project manager must be able to make comparative evaluations of progress in all aspects of the work. Perhaps a few hypothetical examples will help you to better understand the crucial nature of this capability.

Let us say that there are four jobs to be done and let us call them Jobs A, B, C, and D. Assume that Job D cannot be started until the other three have been completed. An analysis of the status of each job reveals:

Job A—Three weeks ahead of schedule
Job B—Eight days behind schedule
Job C—On schedule

There may be many reasons why Jobs A and B are not on schedule, and it is a wise manager who makes a thorough investigation of each instance. But, in the meantime, there is a more crucial issue at hand, and that is bringing the project back on schedule. One immediately apparent alternative is to divert some of the Job A support to Job B. Whether or not this is done, it was the comparative analysis which revealed this staff reassignment as a possible corrective action.

In order to deal with this problem, most project managers try to develop a graphical representation of the total project and to record the latest estimation of completion time on that graph or chart as either on schedule, ahead of schedule, or behind schedule. When a job is either ahead or behind schedule, it is usually expressed as an

Project Assignment / Status Report

Job Name:_____

Job Number:_____

Employee Name:_____

Employee Title:_____

Assignment Date:_____

Customer Name:_____

Customer Charge Number:_____

	Estimated Completion Date			Status as of Week 1	Status as of Week 2	Status as of Week 3	Status as of Week 4
	Org	Rev #1	Rev #2				
Total Job							
Task 1							
Task 2							
Task 3							
Task 4							
Task 5							

Comments Status Week 1

Comments Status Week 2

Comments Status Week 3

Comments Status Week 4

Figure 11-5 Form for project assignment and status reporting

amount of work such as one "person-day" (the estimated amount of work which one person can accomplish in one day's time).

Later in this chapter, two techniques which lend themselves to this aspect of project control will be discussed. Both Gantt Charts and the Critical Path Method (CPM) are commonly employed by effective project managers.

System Checkpoints and Reviews

In order for a project schedule to be effective, it is necessary to have some procedure for monitoring progress and in a broader sense for encouraging personnel to progress. While techniques such as CPM and Gantt Charts will help the project manager in achieving these goals, they tend to understate group attitudes and cohesiveness, which are necessary to get a job done well and on time. Consequently, most project managers use systems reviews and checkpoints as additional aids in project management.

Systems reviews are formal meetings at which various persons and teams report on progress. These meetings are scheduled frequently (for example, bimonthly) and serve a number of purposes:

1. They emphasize continuous accountability as opposed to project-end accountability.

2. They serve as a vehicle for communications between people and groups working on different parts of a project. Hence, they help to build enthusiasm for the "project" not just the "task."

3. They act as a forum for the discussion of unforeseen problems and mutual concerns.

4. They extend the basis for job evaluation to the entire group rather than restricting it to the project manager.

5. They facilitate a process of continuous reinforcement throughout the project so that as people make positive contributions, they receive the reward of compliments from their colleagues. Should they fail to contribute positively, they also experience statements of displeasure and concern.

In any systems project there are a number of more or less critical points in the process. These points are often referred to as checkpoints since it is necessary to make a rather complete review of the project at these points. Common checkpoints in systems analysis include:

1. The Completed Systems Proposal
2. The Completed Systems Design
3. Pre-Implementation Systems Test
4. Post-Implementation Systems Review

The significance of the checkpoint is in the fact that it emphasizes the need for producing deliverables that can be subjected to extended evaluation and review—frequently beyond the limitations of the systems team. For example, after the systems proposal has been developed the team may want to present its proposal to management and to other nonparticipating users, for their comments and criticisms. Furthermore, it may be desirable to do this prior to actually attempting to design the new system so as to avoid expending resources on something which will change as a result of criticisms and comments.

The use of checkpoints to review deliverables has a second advantage. It tends to open the process of systems analysis to the outside in two important ways. First, it makes evident to management and others that the systems team is dedicated to the solution of the problem and that it is making progress in that direction. Second, it gives others an opportunity to gain a greater understanding of what is involved in performing a comprehensive systems analysis and thereby to feel more comfortable with the prospect of participating themselves in similar projects at some future point in time.

Arguments against the use of checkpoints generally focus upon the loss of time required for an extensive review (frequently two to four weeks). Against the advantages that result, this is an insufficient argument.

GANTT CHARTS

The Gantt Chart is a general form of project management control device. It presents a project overview which is almost immediately understandable to nonsystems personnel; hence, it has great value as a means of informing management of project status.

A Gantt Chart has three primary purposes:

1. It informs the manager of what jobs are assigned and who has been assigned to them.
2. It indicates the estimated dates on which jobs are assumed to start and end, and it represents graphically the estimated duration of the job.
3. It indicates the actual dates on which jobs were started and completed and pictures this information.

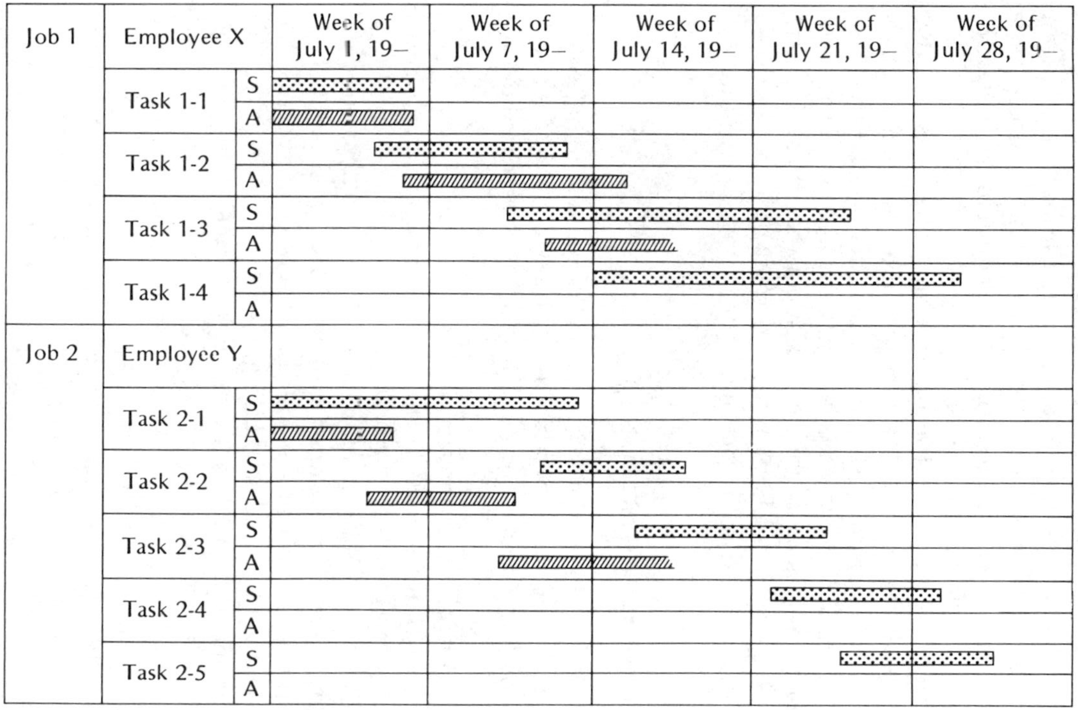

Job 1	Employee X		Week of July 1, 19—	Week of July 7, 19—	Week of July 14, 19—	Week of July 21, 19—	Week of July 28, 19—
	Task 1-1	S	▱▱▱▱▱				
		A	▨▨▨▨▨				
	Task 1-2	S		▱▱▱▱▱▱			
		A		▨▨▨▨▨▨▨			
	Task 1-3	S			▱▱▱▱▱▱▱		
		A			▨▨▨▨		
	Task 1-4	S				▱▱▱▱▱▱▱▱	
		A					
Job 2	Employee Y						
	Task 2-1	S	▱▱▱▱▱▱▱				
		A	▨▨▨▨				
	Task 2-2	S		▱▱▱▱▱			
		A	▨▨▨▨				
	Task 2-3	S			▱▱▱▱▱▱		
		A		▨▨▨▨▨			
	Task 2-4	S				▱▱▱▱	
		A					
	Task 2-5	S				▱▱▱▱▱	
		A					

Figure 11-6 Gantt chart

Like many other management tools, Gantt Charts provide the manager with an early warning if some jobs will not be completed on schedule and/or if others are ahead of schedule. The dark bar (S) in figure 11-6 means estimated schedule, and the light bar (A) means actual progress. Gantt Charts are also helpful in that they present immediate feedback regarding estimates of personnel skill and job complexity. In particular patterns of work emerge quite graphically. Personnel weakness or strength in dealing with certain kinds of job assignments frequently become apparent.

One of the major limitations of Gantt Charts as a management tool is that they do not clearly point out dependencies between jobs. One cannot assume, simply because one job starts later than another, that the latter job is dependent upon the former job. A dependency relationship may not exist between them. If the manager is to keep track of job relationships, some other scheduling method will have to be used in conjunction with the Gantt Chart.

THE CRITICAL PATH METHOD

The Critical Path Method, or CPM, is one method of constructing and analyzing the network of activities which make up any project, be it building a house or installing a new business data processing system.

By carefully constructing a CPM diagram of activities, such as that shown in figure 11-7, the project manager can obtain valuable assistance in three important ways.

First, the CPM diagram points out the logical order in which the various jobs or tasks must be perfor ed. In this sense the CPM diagram is a detailed schedule or time-table of events. Thus, CPM can lead to improved scheduling.

Secondly, the CPM diagram is a means of monitoring project progress since it is possible to compare actual to estimated times on a CPM diagram. In this sense it is a project control device which helps in measuring performance.

Finally, the CPM diagram acts as an early-warning device when difficulties threaten to extend the time required to complete the project. It also reveals where people and equipment resources are available to assist in the critical or problem activities. Thus, it is designed to enable a project leader to take speedy corrective action.

CPM diagrams may be constructed and updated manually. However, because of the size of many projects, they frequently require a computer. The growing use of CPM for project management and

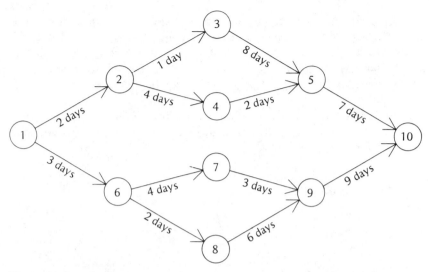

Figure 11-7   CPM network

control has led many computer vendors to provide program packages for CPM or CPM-like capabilities. This discussion is to show you how to construct and maintain a manual CPM network diagram. Before proceeding to the next section, review figure 11-7 once again.

CPM Structural Considerations

A CPM network diagram is made up of a number of "activities," each of which represents a single job or task. Each activity has three parts—a beginning, a duration, and an end.

A CPM Activity

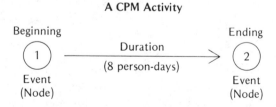

The beginning and ending points (nodes) are called events. An activity is frequently identified by the numbers of its starting and ending points (for example, activity (1,2) is diagrammed above).

The duration of an activity may be expressed in any unit of time or workload. Person-day is the most common workload unit used in systems analysis. The time or workload unit which appears on the arrow, not the length of the arrow, indicates the duration of the activity. (The arrow may be whatever length is convenient for diagramming).

A node may be either a starting event or an ending event. Often, it is both the ending event for one activity and the starting event for the next activity:

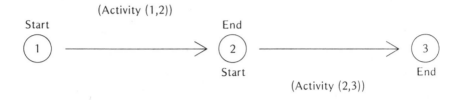

When a number of activities terminate at a single event, all of them must be completed before any new activity can be started at that event. In the following example, activity (4,5) cannot be started until activities (1,4), (2,4), and (3,4) have been completed.

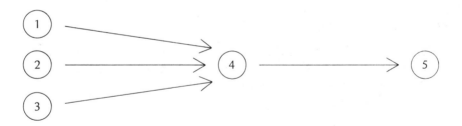

As mentioned earlier, each activity in a CPM diagram is identified by its starting and ending events. So that each activity is unique, no activities can be assigned the same starting and ending points. When two or more activities start and end at the same point, a dummy activity is used. Dummy activities have no duration. They simply enable us to assign different event numbers to activities which start and end at the same point. In the following example, activity (2,3) is a dummy activity which is identified by the broken-lined arrow connecting its starting and ending events:

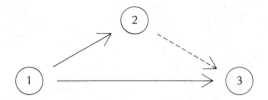

The choice of numbers for events is rather arbitrary; however, for readability, the beginning event number of an activity should be lower than the ending event number of that activity. Review figure 11-7 once again and note the manner in which the various events were numbered to conform to this rule.

Planning Tables and the CPM Diagram

You will recall from our previous discussions in task analysis that it is possible to develop a planning table which clearly indicates the order in which tasks (or, as we now call them, activities) occur. By extending the planning table, as we have in step 1 of figure 11-8, to provide room for a two-digit CPM name and by eliminating the concurrent-activity column, the table becomes the basis for a CPM network diagram. In reviewing step 1 of figure 11-8, note the following:

1. The number of the starting event for any activity is the same as the ending event number for *all* activities which directly precede it.

2. The number of the ending event for any activity is the same as the starting event number for *all* activities which directly follow it.

3. All activities which are not preceded by any other activity start at event 1.

4. All activities which are not followed by any activity end at the *same* last event.

Once the planning table has been completed, it is simply a matter of transferring the information from the table to the diagram, as illustrated in step 2 of figure 11-8.

At this point, a complete picture of the logical order of activities has been developed. However, before the CPM diagram can be a truly useful management tool, we must add more information about the systems project to the diagram.

activity name	directly preceeding	directly following	CPM name
A	—	B, C	1, 2
B	A	D	2, 3
C	A	E	2, 4
D	B	F	3, 5
E	C	F	4, 5
F	D, E	—	5, 6

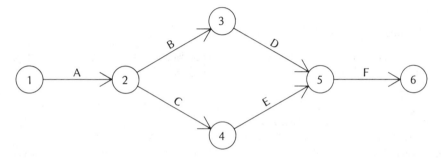

Figure 11-8   CPM network and table

Developing a Network Computation Table

It is at this point that we begin to use earlier estimates of required completion time for each activity. Information from the previously developed planning table is entered into a network computation table like that shown in figure 11-9.

The estimated duration for each activity is also added to that table. A number of values, namely, Earliest Start Time (EST), Earliest Finish Time (EFT), Latest Start Time (LST), Latest Finish Time (LFT), and Slack Time (S), are computed and entered in the table and on the CPM diagram as explained below. These examples, unless otherwise stated, refer to the data contained in figure 11-9.

Earliest Start/Earliest Finish Time

As a project manager, you should know the earliest possible day on which any given activity can begin, assuming that all estimates are correct. The Earliest Start Time (EST) column on the network computation table will contain this information.

Network Table

Activity	Estimated Duration	EST	EFT	LST	LFT	S
1,2	2 days	0	2	2	4	2
2,3	1 day	2	3	4	5	2
2,4	4 days	2	6	7	11	5
3,5	8 days	3	11	5	13	2
4,5	2 days	6	8	11	13	5
5,10	7 days	11	18	13	20	2
1,6	3 days	0	3	0	3	0
6,7	4 days	3	7	4	8	1
6,8	2 days	3	5	3	5	0
7,9	3 days	7	10	8	11	1
8,9	6 days	5	11	5	11	0
9,10	9 days	11	20	11	20	0

Figure 11-9   CPM network table

The Earliest Finish Time (EFT) column will show the earliest possible day on which the activity can finish. The Earliest Finish Time (EFT) is quite easy to compute. We simply add the estimated duration of an activity to the Earliest Start Time (EST).

$$EFT = EST + Duration$$

Notice in each of the cases in figure 11-9 that the EFT is simply the EST + Duration.

The Earliest Start Time is a little more difficult to compute. The EST for the first activity, or activities, is always zero since it can begin on the morning of the first day. The Earliest Start Time of any subsequent activity is always the latest Earliest Finish Time of the activities directly preceding it.

For example, activity (5,10) is directly preceded by two activities, (3,5) and (4,5). Because activity (5,10) cannot begin until all of the activities preceding it are completed, it cannot begin until both (4,5) and (3,5) are completed. Hence, the Earliest Start Time for activity (5,10) will be day 11, which is the latest EFT directly preceding it.

$$EST = Latest\ EFT\ of\ all\ activities\ directly\ preceding\ it$$

Notice that the latest EFT for all activities in figure 11-9 is 20 days. Hence, the earliest possible time for completing the entire project is 20 days. However, not all of the activity progressions within the network require 20 days. This means we could have started some activities later or taken longer than we estimated.

Without increasing the total project completion time beyond the minimum 20 days, let us determine where we could, if necessary, modify our schedule by either starting later or taking longer.

Latest Start/Latest Finish Times

Latest Start Time (LST) represents the latest possible time which an activity can start without delaying the earliest project completion time.

Latest Finish Time (LFT) represents the latest possible time which an activity can be completed without delaying the earliest project completion time.

To complete the LST and LFT computations, we must start with the last activities and work our way backward. Remember, we do not want to increase the total project completion time. Therefore, the Latest Finish Time, like the Earliest Finish Time, will be 20 days for this entire project.

In this case we know that the last activities are activities (5,10) and (9,10). Therefore, both have an LFT of 20 days.

To compute the Latest Start Time for an activity we subtract the estimated duration of the activity from the Latest Finish Time for that activity. Note in figure 11-9 that it is always the case that:

$$\text{LST} = \text{LFT} - \text{Estimated Duration}$$

We have already mentioned that the Latest Finish Time for the last activity is the project completion time, 20 days. Computing the Latest Finish Time for other activities is more complex. When an activity is followed by two or more activities, it must be completed before any of those activities begins. Therefore, the Latest Finish Time must be equal to the earliest of the Latest Start Times which follow it. For example, activity (1,2) is followed by activities (2,3) and (2,4). The Latest Finish Time for activity (1,2) is equal to the earliest of the two LST's which follow it, since it must be completed before they begin.

Once the LST and LFT computations are completed, they may be transferred from the network computation table onto the CPM diagram, as illustrated in figure 11-10. Note that we are now in a position to make several important judgments about this project and about the specific jobs or activities. For one thing we can begin to see which jobs are most critical to meeting our completion-day schedule.

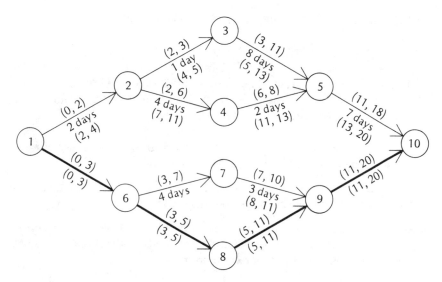

Figure 11-10   CPM network showing latest-start-time and latest-finish-time computations

To emphasize this point, let us return to figure 11-9 and examine the concept of *slack,* or *float,* time.

Slack Time and the Critical Path

Stated simply, Slack time is the extra time which can be consumed by an activity if necessary without affecting one's ability to meet an overall project deadline. Slack Time may be computed by subtracting the Earliest Finish Time from the Latest Finish Time. In other words:

$$\text{Slack Time} = \text{LFT} - \text{EFT}$$

Notice that for some activities the slack time is 0 days. This means that there is absolutely no excessive time, no leeway for getting these jobs or activities done. These are critical activities. They must start and end at the scheduled times or the project completion time will be changed. The progression of critical activities is called the critical path of activities. It is usually represented by a bold line on the CPM diagram, as illustrated in figure 11-10.

If the project completion time is to be shortened, this sequence of critical activities must be completed ahead of schedule. If completion of any of these activities is delayed, the project completion time will be lengthened.

PROGRAM EVALUATION AND REVIEW TECHNIQUE (PERT)

The Program Evaluation and Review Technique (PERT) is another network analysis technique. PERT may be viewed as an extension of the CPM concept, particularly as it relates to time estimates. That is, rather than a single estimate for an event's duration, three different time estimates are made, as follows:

- An optimistic time estimate—the shortest amount of time believed to be needed under the best possible conditions

- A pessimistic time estimate—the longest amount of time believed to be needed under the worst possible conditions

- A most likely time estimate—an amount of time falling some-where between the optimistic and pessimistic estimates, although not necessarily midway between them

Normally, these three estimates are reduced to a single time estimate by using a weighted average based on their probability of proving correct. For example, let us assume that the probability of either our pessimistic or optimistic estimate proving correct is about one in six and that the probability of our most likely estimate proving correct is four out of six. The average or mean time estimate for each event would then be found by using the following formula:

$$AT = \frac{1}{6} \ [OE + (4 \times MLE) + PE]$$

where

$$AT = \text{average time}$$
$$OE = \text{optimistic estimate}$$
$$MLE = \text{most likely estimate}$$
$$PE = \text{pessimistic estimate}$$

We would use this mean estimate in place of the previously described CPM estimate of event duration.

A second extension of the CPM concept frequently associated with PERT is the inclusion of cost data. Although there is no reason why hours of effort cannot be translated into cost data using the conventional CPM, the refinement of time estimates provided through PERT establishes a better basis for estimation and hence tends to encourage the use of cost data.

CHAPTER SUMMARY

A project manager has overall responsibility for the completion of a systems project from problem definition to implementation of the final system. To be effective the project manager must be skilled in (1) scheduling and allocating human and equipment resources, (2) controlling and monitoring project progress, and (3) initiating corrective action as required.

Through task analysis the project manager details the various project jobs into a series of logically dependent steps. By developing a series of standards and applying them to people and jobs, the manager is able to estimate the amount of time required for project completion.

Project Control is an important part of the overall process of project management. Through the use of tools such as Project Status Reports, System Checkpoints, Gantt Charts, CPM, and PERT, the project manager becomes more effective in monitoring progress, detecting potential completion problems, and initiating required corrective action.

WORD LIST

critical activity—an activity for which the slack time is 0 days

critical path— the sequence or progression of critical activities for any given project

critical path method (CPM)—a method of constructing and analyzing the network of activities which make up a given project

dummy activities—CPM activities which have no duration

Earliest Finish Time (EFT)—the earliest possible day on which any activity can finish

Earliest Start Time (EST)—the earliest possible day on which any activity can begin

Gantt Chart—a management control tool which graphically presents a comparison of estimated schedule versus actual progress for each of the jobs associated with a given project

Latest Finish Time (LFT)—the latest possible day on which any activity can be completed without delaying the earliest project completion time

Latest Start Time (LST)—the latest possible day on which any activity can begin without delaying the earliest project completion time

Program Evaluation and Review Technique (PERT)—an extension of the CPM method of network analysis, using optimistic, pessimistic, and most likely time estimates and encouraging the use of cost data as part of the estimation

project manager—one who has overall responsibility for the completion of a systems project from problem definition to implementation of the final system

slack (float)—the extra time which can be consumed by any activity without delaying the earliest project completion time

task analysis—the process of detailing jobs into a series of logically dependent tasks

QUESTIONS AND PROBLEMS

1. Add the two-digit CPM name to the following Activity Planning Table and construct a CPM network diagram from the table.

Activity Planning Table

Activity	Directly Preceding	Directly Following	CPM Name
A	—	B,C	
B	A	D,E	
C	A	F	
D	B	K	
E	B	L	
F	C	J	
G	—	H	
H	G	I	
I	H	J	
J	F,I	—	
K	D	—	
L	E	—	

2. a) Compute the EST, EFT, LST, LFT, and Slack on the following network computation table.

CPM Name	EST. Dur.	EST	EFT	LST	LFT	Slack
1,2	3 days					
2,3	2 days					
2,4	7 days					
3,8	5 days					
3,9	1 day					
4,7	6 days					
1,5	2 days					
5,6	9 days					
6,7	1 day					
7,10	6 days					
8,10	3 days					
9,10	8 days					

b) Which of the above activities constitute the Critical Path?
c) Draw a final CPM network diagram from the above network computation table.

3. Assume that the events in problem 2a above have the following time estimates:

CPM Name	Optimistic Estimate	Pessimistic Estimate	Most Likely Estimate
1,2	2 days	3 days	2.5 days
2,3	1 day	2 days	1.5 days
2,4	4 days	7 days	6 days
3,8	2 days	5 days	3 days
3,9	5 days	1 day	0.75 day
4,7	3 days	6 days	4 days
1,5	0.75 day	2 days	1.5 days
5,6	6 days	9 days	8 days
6,7	1 day	1 day	1 day
7,10	4 days	6 days	5 days
8,10	1 day	3 days	2 days
9,10	4 days	8 days	7 days

Using the preceding information:

a) Compute the average or mean time for each event.
b) Produce a revised network computation table, using the average estimate.
c) Which of the activities now constitute the Critical Path?
d) Draw a final network diagram from the network computation table.

4. In problem 2a, the systems analyst used very conservative or pessimistic estimates. On the surface, such an approach would appear safe. How might the approach have proved risky if the mean time proved to be the actual time required?

5. Answer whichever of the following best applies to your experience in the study of systems analysis:

a) If your professor gave you a detailed course outline/schedule at the beginning of the semester:
    (1) Construct a Gantt Chart which illustrates both estimated and actual completion dates for each major area of study.
    (2) Perform an in-depth task analysis to determine if an improved outline might be possible. Construct a corresponding Gantt Chart.

b) If your professor did not give you a detailed course outline/schedule at the beginning of the semester:
    (1) Perform a task analysis to determine what topics were covered and in what order they were covered. Also determine the approximate time spent on each topic.
    (2) Construct a Gantt Chart with estimated completion dates to be used by your professor the next time he or she teaches the course.

MINICASE—CONTEXT INTERNATIONAL, LTD.

You have been asked to help manage the programming and implementing of the new financial system. You know the following:

• There are 20 programs, each of which can be estimated as requiring 5 (optimistic), 9 (most likely), or 10 (pessimistic) person-days to complete.
• There are 10 programs, each of which can be estimated as requiring 15, 20, or 26 person-days to complete.

- Five of the easier (shorter) programs must be completed before any of the more difficult programs can be started.
- Three of the more difficult (longer) programs cannot be started until all other programs are completed.
- You want a stable work force (you cannot add people, and you will permanently lose people if they are unassigned for noticeable periods of time).

What is your best estimate of how many people should be assigned to the project? How will you organize the programming effort? How will you monitor progress and advise management?

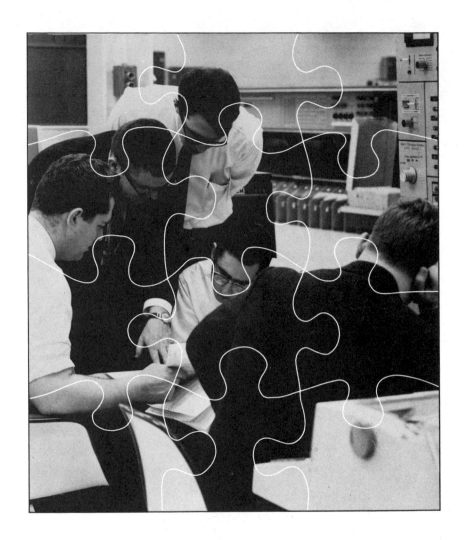

12

Systems Conversion and Implementation

CHAPTER OBJECTIVES

When you finish this chapter you will be able to—

- *Identify and discuss major planning issues affecting systems implementation.*
- *Compare and contrast four different methods of systems conversion providing examples of how each approach is used.*
- *Select an appropriate conversion method given a problem situation.*
- *Discuss systems follow-up procedure.*
- *Discuss and illustrate the "grandparent" system of file security.*
- *Identify and discuss major components of a comprehensive program of systems quality assurance.*

Putting a new system into operation is usually complicated by the fact that there is an older system already in operation. In these instances the analyst has to deal with changing from something familiar to something new and different, while also attending to the mechanics of implementation. Since the concern for simultaneous conversion and implementation is the usual and not the exceptional case, this chapter will address itself to this complicated situation. For readability we will frequently refer to the dual problem as systems implementation.

It is common to find that new systems bring with them new equipment. This may represent a change from manual to automated operation or a change in the level of available machine capacity. In our later more detailed discussion of this matter we will again assume the more complicated case, namely that an equipment replacement will be required as a part of the implementation process.

PLANNING CONSIDERATIONS

Systems implementation is, itself, a systems project of considerable proportion. The individual in charge of this project will have to conduct the implementation with all of the concerns of a good project leader. This means that many of the tools and techniques outlined in the previous chapter will find application during implementation of a new system.

As with most things in systems analysis, good planning can be a decisive factor in determining the ultimate success or failure of systems implementation. Accordingly, some early planning attention should be given to:

1. Assigning systems personnel
2. Training user personnel
3. Structuring user relationships
4. Preparing for new equipment
5. Testing the system

Assignment of Systems Personnel

The project leader cannot assume that because someone was effective in designing a system that he or she should be the logical choice for assisting during the implementation. It is important to assign people to the implementation who have demonstrated ability in dealing with the unique problem situations associated with that process. Let us examine some of the personal qualities which appear most essential to job effectiveness during implementation:

1. Knowledge of the user's role in operating the new system

2. Sufficient knowledge of the operation of the old system to draw comparisons as appropriate

3. Sufficient knowledge of the new system to effectively explain the consequences of user actions

4. Ability to clearly explain to nontechnical personnel how the system operates

5. Effectiveness in working with technical and nontechnical people in problem situations

6. Ability to make sound judgements "under fire"

Systems people who can meet the above criteria are essential if the new system is to be implemented with a minimum of difficulties.

Training of User Personnel

Planning for the formal training of user personnel in the operation of the new system is important. A new system may drastically affect people's lives by changing their work methods, work style, and even their on-the-job associations with other employees. One of the most

Analyst training user on a system which includes a CRT terminal

effective ways of dealing with the potential impact of these changes is to provide a well-designed program of training. This will introduce the user to the new system on a gradual basis before actual implementation. Thus the user will have the opportunity to learn and adjust to the new system in a relatively threat-free environment. An effective training program will seek to:

1. Inform the user about the system in general
2. Inform the user about specific operation of the system
3. Give the user some practice in operating the system
4. Provide opportunity for user feedback

Items 3 and 4 above are particularly important to successful implementation. Through practice the user will be able to make a more objective and practical assessment of the system's potential impact. Also, it will provide answers for many of the questions raised by the earlier nonpractice-oriented training sessions.

The opportunity for feedback is valuable; it recognizes that the teacher must also be a learner. Frequently, feedback during training sessions will identify potential problem areas that were not foreseen by systems personnel. Identification of problems at this point in time frequently permits the systems team to make corrections before actual implementation.

There is another important reason for encouraging user feedback. In so doing, the user is given an opportunity to become a participant in the implementation process. If, by the time the system is to be implemented, users feel that it is "their" system, the chances for a smooth and successful implementation are greatly improved. To identify with the system, the user will have to feel knowledgeable about it, skilled in its operation, and involved in the sense of having had the opportunity to affect what is in the system.

Structuring User Relationships

At no other point in the systems process is the demand for rapid response to problems and potential problems so great as during the actual implementation of a new system. It is also true that this is the time when user reactions will be most critical. In anticipation of this, careful planning is required to ensure that problems can be quickly reconciled through the cooperative efforts of systems and user personnel. In order to accomplish this it will be necessary to: to:

- Detect problems early
- Have an effective procedure available for reporting problems
- Have an agreed upon procedure for joint resolution of problems

User personnel should be encouraged to critically evaluate the new system not only during the training sessions, but as an ongoing responsibility. When this kind of approach is sincerely employed, it generally results in early problem detection.

Once the problem is detected it is necessary to have systems assistance readily available to the user. Delays in responding are not only frustrating to the user, but also raise doubts about the ability of the systems department to respond to real needs. Whenever possible, it is wise to specify capable systems personnel as being "on call" to deal directly with user problems. In this kind of situation, the concern which is expressed through the analyst's quick response can be as important as the solution which will be offered at a later time.

Those problems uncovered during implementation are seldom easy to resolve. The decision is frequently not a "fix it or don't fix it" kind of alternative. Compromises may have to be worked out if the implementation process is to continue. By planning for periodic meetings between these user and systems personnel for the duration of the implementation, the machinery for dealing with crucial problems will be there if and when the need arises. Certainly the alternative of waiting until a major problem occurs before bringing people together is not a reasonable approach. If people meet only during a crisis, they cannot expect a very positive encounter.

Preparing for New Equipment

When a new system also means new equipment, implementation becomes more complex. New equipment introduces additional areas of concern, such as:

1. Structuring a relationship with the equipment vendor
2. Preparing a physical site for the new equipment
3. Installing the new equipment and possibly removing old equipment
4. Training personnel to use the new equipment

The relationship with the equipment vendor will be crucial to the successful completion of each of the other items listed above. To develop a clear mutually agreed upon plan which specifies the duties and responsibilities of each party and the procedures for resolving difficulties is essential. Making the new equipment function at its highest level of efficiency in the shortest possible time is a mutual commitment and should be treated accordingly. In this respect, the use of hollow threats that equipment will be removed or the promise of brighter tomorrows are not the keys to problem solution. What is required is a very close involvement of appropriate vendor personnel in all matters affecting the successful installation of the new equipment and the removal of the old equipment.

A vendor representative, as the on-site project leader, and a well-defined procedure for reporting equipment-related difficulties can be of valuable assistance in dealing with routine new equipment problems. In addition, weekly meetings with the vendor representative in charge of the account allow the analyst to openly discuss the progress which is or is not being made.

Planning the location or site for new equipment requires the bringing together of a number of specialists, including vendor field engineering personnel, as well as user, physical planning, and engineering personnel from within your own institution. Methods used for supplying electrical power, air conditioning, fire protection, and security will have a long-term impact upon the effective use of the equipment. To use instant home remedies can only create instant difficulties for the future. In any event, no new equipment should be brought in until the entire site is completely prepared. Attempting to operate equipment under less than specified conditions can weaken component parts of the equipment causing a high rate of failure later on.

The placement of various pieces of equipment is important to the flow of work. The use of flowcharts which depict the flow of work in relation to the physical layout of the equipment (see chapter 5) can be helpful in arriving at an equipment layout which lends itself to the work situation.

It is usually necessary to have some overlap between the time when new equipment is hooked up and the old equipment is removed. Before the old equipment can be removed it is necessary to certify that:

- The new hardware is reliable.
- The new software is reliable.
- The new system as a whole is able to perform at an acceptable level of efficiency.

In order to make these judgements it is necessary to maintain complete records of all systems problems. It is also well to perform your own test of systems hardware and software and to establish with the vendor beforehand what will be the basis for your acceptance of the system as installed (for example, one month of operation with less than 5 percent total time-loss because of hardware/ software difficulties).

Because it is frequently difficult to pinpoint the exact day on which new equipment can be accepted, some flexibility will be required in scheduling the removal of old equipment. When the vendor supplying the old equipment is a competitor of the new equipment vendor, some tact is required in handling the matter. By approaching the matter honestly and with the expectation that there will be cooperation, few problems generally develop. Respect for the fact that a vendor requires reasonable advanced notice to schedule people and equipment to perform the removal of old equipment is all that most vendors will ask of you.

Plans should allow some training of systems, programming, and equipment operations personnel prior to the delivery of new equipment. Through classroom instruction and field visits to other sites where the same or similar equipment is already operational, the staff can be informed and psychologically ready when the new equipment arrives. Otherwise, personnel may tend to needlessly blame the equipment for their own lack of skill in its use. Knowledgeable personnel will be better equipped to detect genuine machine problems if and when they occur.

Systems Testing

Systems testing is not to be confused with the final acceptance testing of new equipment, which was discussed in the previous section. Rather, it is testing the total new business system prior to attempting its implementation. In effect, it is the point at which segments of the developed system are brought together and tested as a complete system. There are several aspects of systems testing which must be planned for:

- Establishment of testing criteria
- Performing the test
- Evaluation of test results
- "Go" or "no go" determination

There are two methods or sets of criteria for testing a system. The first of these employs the use of "live," or actual, data. The alternative method requires the development of hypothetical, or test, data. It is generally agreed that using test data is the most desirable. Test data can be developed which will test for all conditions of a system, while live data frequently tests for only a limited number of conditions. Furthermore, test data is developed with an understanding of what should be the outcome of processing. Thus, errors can be readily detected.

The reason why many people still prefer to use "live" data is really quite simple. It takes considerable time and effort to develop comprehensive test data. Live data, on the other hand, is usually available at no cost and with little extra effort.

The actual systems test should be conducted in a real-life situation in the sense that personnel who will operate the system later on are involved in the test. Also, every attempt should be made to simulate real work conditions.

The evaluation of test results should be thorough. If there are areas where results cannot be fully explained, then some further testing is warranted. The results of the systems test should be shared with the user prior to drawing any final conclusions.

The decision to proceed with implementation of the system must be a joint decision between systems and user personnel. It must be based upon available fact, such as the systems test results and the skill levels of people who will operate the system. If these facts do not clearly support a "go" decision, then it is wise to plan what must be done before a second systems test can be scheduled. The temptation to plan corrective action and then proceed to implementation without performing a second systems test should be resisted. Within the context of a total system, changes to correct one problem frequently create others. Thorough systems testing will detect these complications.

SELECTING A CONVERSION METHOD

Whenever the implementation of a new system is complicated by the fact that an old system is already in operation, some attention must be given to the way in which the conversion or change will be conducted. There are several commonly used methods for effecting this change-over, including:

1. Parallel systems method
2. Dual systems method
3. Inventory method
4. Pilot systems method

Each of the above methods has its own set of advantages and disadvantages. The particular alternative which is selected will largely depend upon which method systems and user personnel determine will best meet their mutual needs.

Parallel Systems Method

The parallel systems method of conversion involves, as its name would imply, the simultaneous operation of both the old and the new system until the new system can be judged to work effectively. The parallel systems concept is illustrated in figure 12-1.

It can be seen from this illustration that the key to deciding when to go with the new system is the comparative analysis between old and new systems results. In a sense, the old system is the basis for evaluating the new system. Those who argue against the use of a parallel system approach frequently attack the credibility of basing

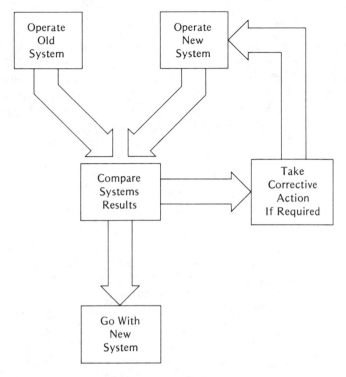

Figure 12-1 Parallel systems method

the evaluation of a new system on the results of a system which does not, itself, do the job adequately.

A second concern regarding the use of parallel systems has to do with the cost of operating two systems simultaneously. Some argue that the potential benefits of such an approach simply aren't worth the considerable expense involved. Others point out that continued operation of the old system takes away resources which are needed during the initial stages of new systems implementation.

On the positive side, the parallel systems approach does provide some assurance that if the new system does not function properly, business operations will continue uninterrupted. Also, there is the fact (however difficult to substantiate) that people seem to feel more secure in not being wholly dependent upon a new and different way of doing business. While it may be a "security blanket" of sorts, it does appear to assist in removing many of the threats associated with change.

Dual Systems Method

The dual systems method gradually phases out the old system while phasing in the new one. For this reason it is often referred to as the "phasing method" or the "gradual change method." We have used the name dual systems method to emphasize that during the conversion some work is being processed on each of two different systems. Figure 12-2 illustrates the dual systems concept.

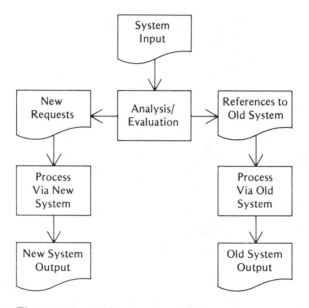

Figure 12-2 Dual systems method

The dual systems approach requires that, at some point, the new system begins to process all new requests and transactions. At the same time, references to materials produced by the old system continue to be processed through that system. In time a washing out of the old system occurs.

The cost associated with the dual systems approach is not very high, since there is little or no direct duplication of work, as in the parallel systems approach. Rather, part of the work is being done on the old system (cleaning up the old work), while the remainder is being committed to the new system.

A second positive feature of the dual systems approach is that it avoids both comparisons and interaction between the old and new systems. By not processing old system workloads, which may contain the inaccuracies of that system, and by not evaluating the new system's performance on the basis of the old, direct assessment of the new system based on its own performance becomes possible.

Critics of the dual systems method argue that it is frequently confusing to both user and customers. User personnel must continually decide which system should produce or has produced a particular document. Furthermore, customers may become confused and concerned if they begin to receive two very different-looking output documents.

A second argument against the dual systems approach centers on the fact that during the period in which both systems are in operation it is difficult to obtain data which presents a total business picture. Because both systems operate and store information differently, obtaining a total is frequently like trying to add apples and pears.

Last, the dual method is frequently criticized for what can be called the "illusion of security." In effect, it does not offer any security because different workloads are being processed by different systems. If the new system were to fail, recovery would require that the user reprocess all of the new system workload through the old system.

Inventory Method

What we call the inventory method is also frequently referred to as the "cold turkey" or "direct" method of systems conversion. The inventory method requires complete, one-time conversion from the old system to the new one, as shown in figure 12-3. In a single large-scale effort, old systems information can be converted for use by the new system. Once this conversion is completed, it is possible to process all information via the new system. In order to conduct this kind of conversion without interrupting the normal business flow, nonbusiness hours (particularly weekends) are generally used to bring the new system up.

Supporters of the inventory method argue that it is a clear, quick, and reasonably inexpensive method for installing a new system. They also point to the fact that the cost of alternative methods is high and their true security value questionable.

Critics of the inventory approach are frequently vehement in their opposition. They see it as risky, if not potentially suicidal. Few opinions about the inventory method, either pro or con, can be called "lukewarm." Rather, the approach almost always polarizes opinion when it is discussed as a conversion alternative.

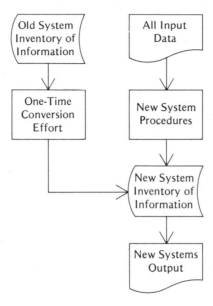

Figure 12-3 Inventory method

Pilot Systems Method

The pilot system requires that only a small portion of the new system be implemented in either a parallel or a dual systems fashion, while the major portion or all of the workload continues to be processed via the old system. Figure 12-4 illustrates both the dual-pilot and parallel-pilot concepts.

An independent analysis and evaluation is made of the pilot system. Based upon the success or failure of the initial pilot system, one of the following approaches will be used:

1. To make required corrections or modifications while continuing the pilot system as is

2. To temporarily revert to the old system while considerable changes are being made to the pilot system

3. To expand the pilot system to accommodate more, but not all, of the workload

4. To convert entirely to the new system using a dual or inventory systems approach

The dual-pilot system shares in many positive and negative criticisms of dual systems with one major exception. That is, only a por-

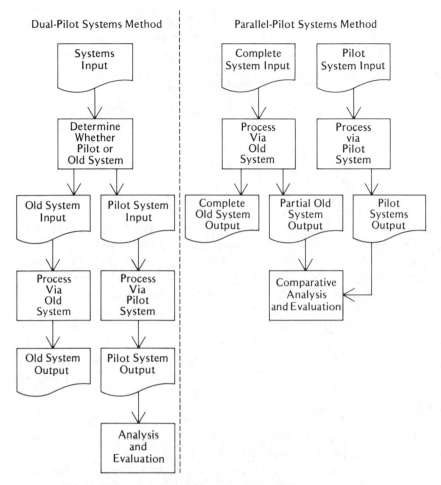

Figure 12-4 Dual-pilot systems method and parallel-pilot systems method

tion of the new system is being installed and this does afford some measure of security in the event that the pilot system does not work.

The parallel-pilot system approach has a strong conceptual similarity to the parallel systems approach described previously. For example, the entire old system continues to operate. Also, the evaluation of the pilot system is based upon its comparison to partial old systems results. Thus, it too shares in the positive and negative criticisms of its conceptual counterpart with one notable exception. That is, the cost is significantly reduced because of limited duplication between the new and old systems.

One general criticism which is frequently launched against both pilot systems approaches is that they frequently do not, because they are partial systems, provide an adequate basis for evaluating the total new system. Therefore, it is important to clearly establish when a pilot system is or is not a good measure for evaluating the total new system. This could be an important factor in the final decision to use, or not to use, the pilot approach.

SYSTEMS FOLLOW-UP

Systems follow-up is a most important aspect of the total systems process and one which is closely associated with the process of systems implementation. It is not enough for the systems analyst to simply throw the switch for a new system, make a few evaluations, and then leave. A period of extended commitment after installation is required, during which the analyst will carefully examine the actual system performance, compare these findings to initial expectations, and make short- and long-term recommendations regarding the system. This most important aspect of the work should be done in close cooperation with both the user(s) and those higher levels of management which have an interest in or a responsibility for the total project.

One approach which is strongly recommended is to schedule a period of time for a formal post-implementation review. To accomplish this a group or task force is established to review the implemented system and determine:

1. Does the system adequately meet the initial goals and objectives set forth in the systems proposal?
2. Do the key systems procedures and operations function smoothly and efficiently?
3. Have the various criteria for systems decision making, security, training, and control been met?
4. Have the projected benefits been realized and the prior liabilities been eliminated?
5. Are there unaccounted for problems which require additional systems development?
6. What are the users' perceptions/reactions?

Normally, after a post-implementation review, the analyst reports on the above to management with recommendations regarding work to be done in order for the system to be considered complete. Also, it is common to spell out specific areas not covered in this project which should be considered for future systems development.

Obtaining a Sign-Off

Bringing the systems project to an official close, while also ensuring that the commitment to the user has been met, is most important. Until the project has been officially closed, key systems personnel are not truly free for reassignment to other areas. It does not take much skill in mathematics to determine the net effect of two or three unfinished projects upon a systems department's capacity for responding to new and crucial problems. Thus, there is the need for firmness as well as fairness in performing systems follow-up. A practice commonly referred to as the "systems sign-off" is designed to provide the required measure of firmness to the follow-up. Briefly, a sign-off is a formal statement by all parties that the system meets initial goals and objectives and that it is, in fact, to be considered operational.

The sign-off statement will also generally contain some reference to the procedures which should be followed in seeking any future changes or enhancements to the system. In effect this means that any request for change will not be automatically honored, but rather that its priority will be evaluated within the context of other systems requests and ongoing projects.

QUALITY ASSURANCE OF NEW SYSTEMS

Delivering a high-quality system has been the focus of discussions throughout this text. However, unless mechanisms and procedures are in place to assure a continuing high quality of performance, the system's longer-term effectiveness will be in jeopardy.

A number of quality assurance procedures and methodologies have been alluded to earlier, particularly in the discussions of systems design. Nonetheless, it is worth recapping them at this point, emphasizing that they must be fully operational and in place at implementation time. The topics to be covered in this section include:

- Security
- Personal privacy
- Environmental control
- Systems backup and recovery
- Test system migration
- Enforcement of systems standards
- Systems internal control, effectiveness, and efficiency reviews

Each of the above topics is an important aspect of any comprehensive program of systems quality assurance.

Security

The topic of security is a broad one. In this section, we will briefly
address:

- Physical security of the computing facility
- Disaster planning
- Securing the batch system
- Securing the online system

During the "riots" of the late 1960s and early 1970s, it became
painfully apparent that many computing facilities were not constructed with potential security problems in mind. Many were placed
in highly visible, easily accessed areas. Many lacked even the most
fundamental mechanisms and procedures needed for protection from
illegal entry, theft of equipment or supplies, sabotage, and various
natural phenomena such as fire or flood. Today, the computing
facility with a comprehensive program of quality assurance has the
necessary fire detection and extinguishing systems, subfloor water
detection equipment, restricted entry (for example, card entry) systems, surveillance equipment (for example, closed-circuit TV monitoring systems), and an action plan for various possible natural and
human-induced disasters.

In recent years, attention has also been paid to achieving a higher
level of security for both batch and online data processing systems.
As institutions have become more highly computerized, the opportunities for computer fraud and the potential negative consequences
of making honest mistakes have expanded.

For batch data processing systems, improved security of the computing facility has also provided added systems security, because
the batch programs and files reside in the computing facility.
Additional steps to limit access to tape and disk pack libraries and
to separate responsibilities for systems development and day-to-day
operations have also improved the situation.

For online systems, the problem is more complex. In an online
environment, it is not necessary to break into the computer facility
to make unauthorized changes. The potential for security problems
is further increased when the system permits "dial-up" access
through conventional telephone service. A succession of widely publicized instances of fraud has caused many organizations to rethink
allowing dial-up access to their systems. Rather, they have adopted
a closed system approach whereby no dial-up access capability
exists. Many organizations have instituted a variety of methods and
techniques to further control terminal access, including the following:

- *Password security*—one or a series of established passwords must be entered by the terminal operator to gain access to the system. Access is usually further restricted to some subset of online transactions identified as being suitable. Passwords should be changed frequently but on an irregular schedule.

- *Physical terminal security*—terminals can be placed in reasonably secure office areas and equipped with features such as keyboard locks. When leaving the terminal, the operator literally locks the keyboard without having to turn the terminal off.

- *Subschema design*—as mentioned earlier, the data base schema can be further broken down into specific subschemas. In so doing, terminal access may be restricted to particular portions of the data base.

- *System monitoring*—terminal activity may be monitored from the system console or a specially designated control terminal. It is not uncommon for large financial institutions to hire qualified personnel to monitor terminal activity explicitly to catch someone in the act of attempting to crack the system.

The development of new security techniques will continue to mushroom as more and more institutions expand their data processing involvement and particularly their online, data base applications.

Personal Privacy

Much publicity has been given to the issue of personal privacy. Our comments on the subject will be brief. The legal implications of maintaining data about people are only beginning to be clarified. However, the privacy issue has aroused enough concern to cause institutions with an effective quality assurance program to:

- Institute procedures for explaining to people what will be done with personal information

- Institute procedures for quickly responding to people's requests for a statement of what data about them is being stored on the computer

- Review proposed systems data requirements to eliminate personal data that are either sensitive or simply inessential to effective performance of the proposed system

- Perform data integrity reviews to ensure that the data being maintained are accurate and reliable

Environmental Controls

With the development of increasingly advanced electronic circuitry, problems caused by the computer room's temperature and humidity have become less acute. It is simply easier and less expensive now to meet environmental requirements. However, a number of other potential problems still warrant attention, such as:

- Dust control—this problem has been exacerbated by extremely high-speed printers and has produced a tendency to place paper- and card-using peripherals in rooms physically separated from the main computer and disk units.

- Media storage environment controls—The importance of effective backup and recovery systems for large-scale data base systems has increased the emphasis on storing tape, disk, and card backup files under dust-free conditions of appropriate temperature and humidity. Also, programs for periodic tape cleaning and certification are frequently established.

Systems Backup and Recovery

Effective backup and recovery methods and procedures are central to any quality assurance program. There are both similarities and differences between the techniques used for backup and recovery of batch and online systems.

The most commonly used system for information backup in a batch environment is called the "grandparent system." Figure 12-5 shows how such a system operates with computer-generated tape and card files (a similar process would be followed if disk files were involved).

In an online environment, the process works somewhat differently. Imagine for a moment a large centrally stored data base against which online transactions are being processed by 150 remote terminal operators. A number of actions occur simultaneously with the transactional processing to establish backup and facilitate recovery, including:

1. Transactional "logging"—as online transactions are processed against the data base, they are logged onto an independent file called a transaction "log" or "audit" file. The transaction log is generally a tape file.

2. DBMS before-and-after looks—for transactions that modify the data base, the DBMS may take before-and-after "snapshots" of the data base areas being changed. If problems arise, the DBMS can establish the last-known correct image of the data base.

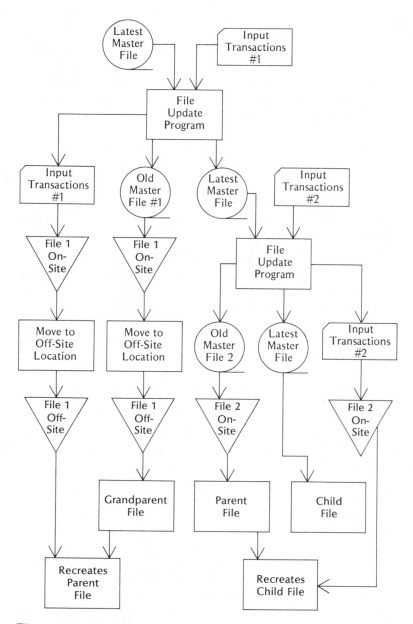

Figure 12-5 Grandparent system of file security

3. Periodic checkpoints—these are indicators on the transactional log tape or look files that can be used as restart points
   if a problem arises. They enable recovery to occur without
   requiring that all of the day's transactions be reprocessed.

Besides the above actions, it is usual at the close of each day to "dump" the entire data base onto magnetic tape. The tape's use in the recovery process will become apparent shortly.

The actual recovery process resembles that used for batch systems, because it is a batch process itself. Following are the steps required for recovery:

- If a minor failure occurs, the DBMS may be able to back out the bad transaction, using its before-and-after look files, and recover the system.

- If a more serious failure occurs, it may be necessary to return to the last checkpoint and reprocess in batch form all the subsequent transactions from the log tape.

- If a very serious failure occurs (for example, loss of the total data base), it will be necessary to first reconstruct the data base, using the tape dump from the previous night. After this step, the transaction log tape will be batch processed against the data base. It is worth noting that transactions collected over hours of online processing can be batch processed in a relatively short time by the recovery programs.

A grandparent filing system is also employed with a data base dump and transactional log tapes. Figure 12-6 illustrates the long recovery process.

All quality data base management systems come equipped with backup and recovery capabilities. One should avoid commitment to a DBMS that does not come equipped with these capabilities.

Test System Migration

We would all like to think that the delivery of a finalized system means that the system will operate without problems and without required changes. However, the complexities of large-scale systems and human frailty combine to frustrate this hope. For that reason, it is important first to construct a proper testing environment and procedures for making necessary changes during the development of a new system. Also, added safeguards are required to safely make changes to the new system once it is operating. In particular, it is not wise to insert changes in the production version of the system until the test results have been thoroughly reviewed and their accuracy has been verified. Also, if problems occur in implementing the changes, quick reconstruction of all systems components to their original state must be possible.

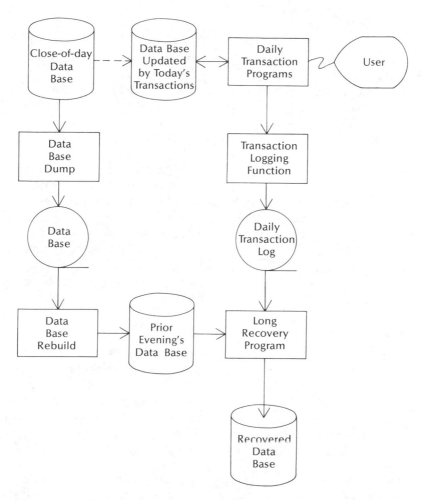

Figure 12-6   Data base backup and long recovery process

Normally, it is expected that a separate test data base will be maintained and that test version copies of the production programs will be established. Programmers are generally required to use this test library of programs and files. Several steps should be performed once the programmer has obtained "test" results clearly indicating that the changes to the test system have been properly made.

First, an analyst and other programmers should review the altered program and test results, using a "code review" or structured walkthrough procedure similar to that used during initial development. If logic problems are discovered or the program changes have not been performed according to existing standards, the programmer must make the additional changes, rerun the test, and attend another code review. Only when the reviewers agree that the revisions are in final form does the implementation of the changes move forward. At times, people may want to shortcut this process, because it may seem that just a few more minor corrections remain to be made. However, changes that are not thoroughly reviewed and tested have a very high failure rate. Remember too that, when an online production system fails, it frequently causes sizable losses by idling the time of people who need the system to efficiently process their work.

Once all clearances have been obtained for the changes, implementation and live system testing are normally scheduled for times when little or no service interruption will occur. Also, it is wise to schedule changes so as to leave enough time to revert to the old system if additional problems occur during implementation.

By following the above practices, the organization is assured of an effective migration from the test to the production environment when a new system is being implemented or when system changes or enhancements become necessary.

Enforcement of System Standards

This text presents the many legitimate options for performing systems development. A given organization may seek greater conformity and control by adopting some subset of all options as the system development standards it expects its personnel to use. Thus, one organization may adopt HIPO for its staff to use, while another may adopt data flow diagrams. (Still another may expect HIPO to be used for certain aspects of the work and data flow diagrams for other aspects.)

Similar sets of standards are generally established for programming and computer operations. These standards help to guide people in understanding what they are to do, how to do it, and what evidence (documentation) to produce for their work to be considered complete.

Enforcing standards has traditionally received less attention than developing them has. Thus, in many instances tremendous effort is devoted to developing standards that are later only marginally adhered to, if at all. The reasons for this outcome include:

- A nonconformist attitude that seemed legitimate, even desirable, during the early days of computing
- Overelaboration of standards
- A lack of management awareness of how to enforce standards in a technical environment
- A concern for the cost and/or time required to enforce standards

Today there is a better understanding of the need to stress professionalism (which involves conforming to sound work practices) instead of assuming that individual preference is somehow or other always the "best" way. Performing work according to generally accepted practices helps the individual grow as a professional. This approach also provides assurance that the work is being properly performed and can be understood by others if the need arises through employee turnover.

Organizations have begun to back away from overly restrictive and elaborate standards, thus making them more palatable to intelligent and creative people. In attempting to develop standards today, the wise person recognizes the inverse relationship between the size of the standards manual and the probability that the manual will ever be read.

Finally, there is a growing awareness that standards can be enforced only by carefully reviewing the ongoing work for its conformity to standards. Whether this enforcement takes the form of a review panel (for example, the program code review described earlier) or a supervisory review of a person's work is less important than the fact that a review is performed. We have noted on several occasions that project leaders used to be selected almost solely because of their previous technical achievement and that they frequently continued to spend large amounts of time writing computer programs. In recent years, the profession has changed dramatically in these respects. People are increasingly selected for project leadership because of their demonstrated managerial ability. Also, they are expected to supervise and manage others' work rather than simply do it themselves.

Systems Internal Control, Effectiveness, and Efficiency Reviews

An organization wants absolute assurance that its most sensitive systems, such as financial and accounting systems, have sufficient built-in controls. Internal and external EDP auditors frequently perform internal control reviews and thereby provide the higher level of assurance required for these kinds of systems.

As organizations have become more heavily computerized, the EDP budget for people and equipment has grown dramatically. Hence, an increasing number of organizations periodically use external reviewers to perform effectiveness and efficiency studies of their EDP installations and major systems applications. At first, many EDP people viewed these reviews negatively. The fact that many of these early reviews focused solely on problems did not help the situation.

Today there is greater recognition that EDP organizations are, like any other organization, subject to problems of inefficiency and reduced effectiveness. A properly conducted efficiency and effectiveness review can help an EDP organization better understand its strengths and weaknesses.

CHAPTER SUMMARY

The process of implementing a new system is frequently complicated by the fact that there is an older system already in operation. At such times it is necessary to deal simultaneously with the dynamics of both systems implementation and systems conversion.

Effective systems implementation requires that the analyst have a plan ready which will deal with a number of important issues including: (1) assignment of systems personnel to the implementation; (2) training of user personnel; (3) structuring of user relationships; (4) preparation for new equipment; and (5) testing of the new system.

There are a number of commonly used methods for converting from an existing system to a new one, including the parallel systems, dual systems, inventory, and pilot system approaches. Each has advantages and disadvantages, and selection of the proper method depends upon existing conditions and circumstances.

A period of extended commitment after implementation is required. During this follow-up period the new system is fully evaluated, required changes are made, and the systems project is brought to a formal close.

For a new system to have longer-term effectiveness, a well-established quality assurance program must be introduced at the time of installation. This quality assurance program should address a broad range of issues, including security, personal privacy, environmental control, systems backup and recovery, test system migration, enforcement of systems standards, and systems internal control, effectiveness, and efficiency reviews.

WORD LIST

data base recovery—a procedure that follows systems failure and uses transaction log and backup files to rebuild the data base to its most current form, just as it was before the systems failure

dual systems—a method of systems conversion requiring that the old system be gradually phased out as the new system is gradually phased in

grandparent system—a commonly used method of providing information backup to a system

inventory method—a method of systems conversion commonly referred to as the "direct" or "cold turkey" approach which requires a complete, one-time conversion from the old to the new system

parallel systems—a method of systems conversion requiring the simultaneous operation of both the new and the old system until such time as the new system is judged to work effectively

pilot systems—a method of systems conversion which requires that a small portion of the new system be implemented and evaluated while the major portion of the workload continues to be processed via the old system

systems conversion—the process of changing over from an old system to a new system

systems implementation—the process of placing a new system into operation

systems sign off—a formal statement agreed to by all parties involved in the systems project which attests to the fact that the system meets initial goals and objectives and that it is furthermore considered to be fully operational

transactional logging—as online transactions are processed against the data base, they are posted to an independent file called a transaction "log" or "audit" file

QUESTIONS AND PROBLEMS

1. Select a systems conversion method to be used in each of the following instances. State the rationale for your selection.

   Situation 1:
   Bogard Industries plans to implement a new, comprehensive financial management system. The current system, while it is judged to be both accurate and reliable, is very limited in the manner in which it is able to manipulate and report on basic financial data. Good financial management is the cornerstone of Bogard's reputation, and the company enjoys a healthy financial posture which it attributes to this management ability.

   Situation 2:
   Questo Mining Corp. plans to implement a new accounts receivable subsystem. The current system was described by a group of internal auditors as "an antiquated, ineffective operation riddled with inaccuracies and inconsistencies." In the pre-

liminary systems study which followed that audit, it was estimated that systems problems in the accounts receivable area had cost the company approximately $20 million over the last 3–5 years.

2. Your manager has asked you to develop a presentation on quality assurance of new systems, which you will deliver to a group of newly hired trainees. Prepare a topical outline with accompanying illustrative examples.

3. Jerry Teabon, an analyst with E. F. Whitney, has recently been criticized by a number of users. They argue that Teabon has given them inadequate training for operating the new order processing/inventory management system.

   When queried by his boss, Stan Mather, Jerry points out that he is the only qualified trainer and that literally hundreds of people must be trained. He is quite annoyed by the complaints, because he has worked hard to give as many people as possible at least a minimal amount of training.

   If you were in Mather's place, how would you respond, and why?

4. The new production planning system at Squires International has been operating for nearly six weeks without any noticeable difficulties. Today the systems department was notified that data on the file appear to be inaccurate. George Thomas, a senior analyst, troubleshoots the system and quickly isolates the problem.

   George believes the system can be brought back "up" by midafternoon, provided that this action plan is followed:

   • Fix the computer program causing the problem.
   • Offload the bad data and quickly patch them up.
   • Reload the data.
   • Bring the system up and resume using it.

   Any alternative approach will probably keep the system out of use until tomorrow morning. Your manager has asked for your advice.

   a. What are the pitfalls of George's approach?
   b. What alternatives are there?
   c. How would you like to see the matter handled?

5. Present a reasonable approach to be used during systems follow-up which will assist in bringing the systems project to a speedy and effective completion.

MINICASE—CONTEXT INTERNATIONAL, LTD.

You have been asked to lead the planning of the financial system implementation at Context International headquarters and at the firm's eight subsidiaries. The task promises to be a complex one, because:

- Three subsidiaries are essentially manual operations.

  Two of these have reasonably sound, albeit limited, financial systems. One has serious problems.

- Four subsidiaries have considerable, but incomplete, computerized systems.

  The above systems are incompatible and will require file conversion before new systems implementation.

  Two of the above subsidiaries have not fully supported the new systems effort throughout.

- One of the subsidiaries is in deep financial difficulty and has laid off many employees. Also, many other good employees have recently left the firm.

  Both the manual and limited data processing systems still function, but without the knowledgeable people required to prevent unexpected breakdowns.

- The corporate management would like all conversions and implementations to be finished in the next 9–12 months.

In developing your plan, you are to prepare the following:

1. A statement of overall strategy to be followed during implementation, focusing on technical and organizational issues

2. A statement of specific areas of special concern—for example, where you expect the greatest difficulties, and why

3. A concrete proposal for modifying the strategy to maximize success in those areas

4. A proposed implementation methodology (for example, parallel, phased, and so on) for each subsidiary, based on its situation (for example, a manual system with serious problems)

How do you feel about corporate management's level of expectation? What will you do about it?

Although your data are insufficient for projecting actual implementation time requirements, what about relative time estimates (for example, what would you recommend doing first, second, and so on)? Prepare a hypothetical Gantt Chart, illustrating your proposed implementation plan. (Remember to leave time for user training, revisions, and so on.)

Appendix A

Training Program in Effective Communications for Systems Analysts

There is a limited "hard core" of communications experts within most organizations. This is particularly true of systems and data processing departments. Few systems analysts have received even a minimum of formal training in the development of communications skills. This means that there is at least some initial cost to an organization that wants to develop these skills in its systems personnel. However, given the assumption that a system is only as good as the quality of the information upon which it is built, this cost appears to be justifiable.

In developing a training program that emphasizes effective communications for systems analysts, a number of ingredients are fundamental.

- Orientation
- Communications Theory
- Controlled Learning
- On-the-Job Training
- Self-Awareness
- Periodic Seminars

Orientation

Some time should be spent in attempting to orient the analyst to the idea that systems analysis is a communications process. During this orientation, the importance of maintaining a base of accurate and reliable information throughout the systems process should be stressed. It should be pointed out that the systems interview is an important aid in securing information at all levels of encounter. Also, the growing importance of group process skills during the development of large-scale systems should be stressed. Examples should be provided to dramatize what can happen in terms of personal difficulties and cost effectiveness when a systems design is based on erroneous, unreliable, and/or incomplete information.

Communications Theory

A cross-section of interviewing and group process ideas and practices should be provided to the analyst. Special lecturers from a variety of fields such as public relations, data processing, personnel, social welfare, job employment, organizational development, and psychological counseling are good starting points for the training program. Naturally an organization's pocketbook will make a difference in what it can afford to provide to its analysts in the way of lecturers, but the wise manager will not skimp in this area—particularly when he or she is attempting to build in-house expertise.

Learning in a Controlled Environment

During the initial stages of learning, it is best to exercise considerable control over the variables affecting the learner. Therefore, it is desirable to provide close supervision and to restrict the learner's environment as opposed to "setting him loose" on the world-at-large. Two controlled methods are particularly effective in teaching the analyst to communicate, namely, the case-study approach and the role-playing approach.

In the case-study approach the analyst is presented with a variety of studies, each of which dramatizes one or several of the "Do's" and "Don't's" of communications. Since case studies are most helpful when their meaning is clear, they should be kept quite simple. Long and involved case studies which simultaneously reflect all of the complexities of life are generally poor examples. Through the case-study approach the analyst may gain significant insight into how to avoid gross errors.

Supervised role-playing is a more advanced type of controlled learning. In the role-playing situation, although the content or stage for communications is outlined, the personal confrontation is less controlled. That is, traditionally one person is assigned the role of systems analyst, told what he or she is assigned to investigate, and given a general background on the surrounding conditions.

Additional members of the training class are assigned the role of the user personnel and given identities in terms of responsibilities and functions.

During the "play out" period, the actions of all parties are important to the ultimate goal of effective communications. The analyst's role is important in pointing up technical errors and in enabling the analyst to gain familiarity with the process of communication. The user roles are equally important since in portraying other people one projects a personal view of those people. Thus, by introspectively analyzing one's actions while playing a particular role, insight may be gained into the personal biases and prejudices which one harbors.

By using both the case-study and the role-playing approaches to communications training, an optimal opportunity to improve is provided to the systems analyst. Even when only one method can be included in the program, it can usually be assumed that the analyst will enter the field with greater understanding and caution.

On-the-Job Training

Rather than assume that the classroom or even laboratory role-playing provides complete training, it is a good idea to carry the concept of supervised learning into the field. Many organizations will accomplish this by employing something akin to a "big brother" technique, wherein an experienced systems analyst accompanies the novice as a silent partner for some limited period of time. At the conclusion of each field session, the actions of the novice are discussed openly and honestly. The senior partner makes recommendations to the analyst for personal improvement and sets up additional time to cover weak areas. However, an organization must be careful not to over-burden its best personnel with this type of assignment.

Self-Awareness

It has been suggested that a good starting point for dealing with many kinds of problems is to look at oneself as a potential problem source. Any training program of the kind outlined above, if it is to be truly effective, must focus on developing the systems analyst's critical awareness of self. Only if the analyst has a genuine understanding of his or her beliefs, attitudes and hang-ups will it be possible to improve upon the ability to communicate effectively with others. Because people are dynamic and changing, self-awareness is not a one-time excursion. It is a continual process of assessment and reassessment. Certainly, analysts who have experienced professionally conducted self-awareness sessions tend to be more keen in their use of self-criticism. Nevertheless, a continuous program which encourages open and honest exchange among analysts can yield many dividends.

Periodic Seminars

As with all dynamic situations, new problems will continually arise during interpersonal communication for which there are no ready answers. To effectively deal with these problems and to foster the mutual interchange of creative ideas, it is a wise manager who periodically schedules systems seminars specifically dedicated to further development of effective communications ideas and practices.

Appendix B

Data Processing Equipment Costs

Arriving at Data Processing Equipment Costs

A first consideration in any project of this type is to have available a complete and up-to-date inventory report of data processing equipment. The equipment inventory report should be grouped by equipment vendor and function and detailed to include each physical piece of equipment. For example, all IBM 029 keypunch machines should be individually listed under the general heading of keypunch equipment. The reason for grouping equipment by vendor rather than by some other criteria, such as room location, is that monthly charge statements (which should be routinely checked against this report) will be received by the vendor. The reason for grouping equipment by function is apparent from the discussion of the cost center concept found in chapter 7.

When the data processing equipment in question is a computer, it is not sufficient to simply list it as a single item on the inventory report. To the contrary, it is absolutely imperative that the report be detailed to the point of having a separate entry for each computer device. Consequently, a disk drive or a tape unit would appear as a separate entry on the report. Note the kinds of detailed information which appear on the sample report shown in figure B-1. Suggested information on that sample report includes:

- *Serial Number*   This unique number is assigned by the manufacturer or vendor and can generally be obtained from the equipment vendor.
- *Vendor Name*   This is the complete or abbreviated name of the party from whom you are renting or have purchased the equipment, such as Univac, IBM, or Burroughs.
- *Vendor Model Number*   This is the commercial name given by the vendor to the particular unit, such as 029 or B-3501.
- *Description*   This is the functional name of the unit, such as keypunch or 1000 cpm card reader.
- *Function Code*   This code describes the function or use of the equipment. For example: D.P. = data preparation; S.P. = systems and programming; C.P. = computer processing. (This coding will be important to later attempts at cost center development).

Serial Number	Vendor	Model Number	Description	Func. Code	Installed/ Discont.	Gross Cost/Mo.	Percent Discount	Net Cost/Mo.	Yearly Cost	Room Location

Figure B-1 Equipment inventory report

- *Date Installed/Discontinued*   This date reflects the date on which payment for the unit began or ended. It is not necessarily either the date on which delivery was made or the date on which the equipment was hooked up. It is only when the vendor and the data processing center have agreed that the equipment is operable that the equipment can be considered as installed, hence, billable. Similarly, it is the agreed-upon date for discontinuance that should determine when payment legally ceases, not disconnecting or physically removing the equipment.
- *Gross Monthly Cost*   Since most data processing equipment is paid for on a monthly rental or installment basis, it is important to indicate the gross monthly cost for that equipment in the report.
- *Percent Discount*   Many customers, particularly in government and education, receive a discount on the usual market price. However, not all pieces of equipment from a vendor will receive the same rate of discount .
- *Net Monthly Cost*   For purposes of verifying monthly billing statements, it is necessary to show the net monthly cost as well as the gross monthly cost and the percentage of discount. In this way it will be possible to pinpoint billing errors if and when they occur.
- *Yearly Cost*   This is a most helpful cost element to have readily available, particularly when it becomes necessary to plan for equipment changes and modifications.
- *Room Location*   This information is kept primarily to ease the task of verifying that equipment does exist, as is required in most audit situations.

Additional optional cost information which might have been contained in the sample report include:

1. Costs which were not paid to the vendor because of excessive "down time" or machine malfunctions.
2. Projected annual costs for equipment not installed for a full year.
3. Maintenance costs for services obtained either through separate contract or on a per call basis.
4. Costs for specific software packages including operating systems which have been purchased or leased.

Once inventory/cost information has been compiled, the systems analyst is in a sound position to begin an analysis of data processing service costs as they apply to particular jobs.

Equipment Utilization

The extent to which equipment is used will have a significant impact upon the actual cost of services provided. This is particularly true when there is a fixed monthly rental charge associated with a piece of equipment. In other words, the cost per unit of time can decrease as the total amount of equipment utilization increases. Conversely, the cost per unit of time can increase as the equipment utilization decreases. Using an oversimplified example of this cost/utilization relationship, the following attempts to show the per hour cost of using equipment with a fixed rental charge of $5,000/month:

Low Utilization Example:

$$\frac{\$5,000/\text{month rental}}{50 \text{ hrs./month utilization}} = \$100/\text{hour average cost/hour}$$

High Utilization Example:

$$\frac{\$5,000/\text{month rental}}{250 \text{ hrs./month utilization}} = \$20/\text{hour average cost/hour}$$

In practice, utilization data is frequently both difficult to capture and difficult to identify with a particular job. To effectively determine who is using what piece of equipment and for how long will require a well-structured system for equipment control.

Equipment Control—Customer Jobs

If utilization data is to be meaningful, it must be accumulated by both customer and job. In this way it is possible to determine both specific job costs and total equipment allocation costs for a particular customer. Control by customer is frequently accomplished through the use of customer charge numbers. Figure B-2 is a sample list of charge numbers for a typical college or university. If at all possible, it is wise to adopt a charge number system which exists rather than add to the growing proliferation of numbering systems.

In addition to establishing a list of customer charge numbers, it is necessary to develop forms and procedures which will insure that the proper charge number will be associated with the proper job. This is generally accomplished by creating "job request sheets" which are prepared by the customer. All "job request sheets" should have a place for customer charge number and job description. Furthermore, it is important that equipment utilization, job, and customer charge number be associated at the time that the equipment is being used (see figure B-3 for an example of a combined job request and machine utilization form).

INSTRUCTION & DEPARTMENTAL RESEARCH

40000 Agriculture, Division of
40001 Chairman's Office
40003 Agricultural Business
40004 Agricultural Engineering
40005 Agronomy
40006 Animal Husbandry
40022 Dairy Technology
40071 Ornamental Horticulture
40090 Ag.-Vocational Training

44000 Business, Division of
44001 Chairman's Office
44002 Accounting
44014 Business Administration
44024 Data Processing
44087 Secretarial Science
44090 Bus.-Vocational Training

53000 Food Service Administration, Division of
53001 Chairman's Office
53035 Food Service Administration
53090 Food-Vocational Training

54000 General Education, Division of
54001 Chairman's Office
54013 Biological Sciences
54045 Health and Physical Education
54049 Humanities
54063 Mathematical and Physical Sciences
54086 Science Laboratory Technology
54089 Social Sciences

71000 Nursery Education, Division of
71001 Chairman's Office
71068 Nursery Education

86000 Other
86005 Evening Division: Chairman's Office
86010 Evening Division: Instruction
86020 Summer Sessions: Chairman's Office
86025 Summer Sessions: Instruction
86031 Educational Communications
86035 Liberal Arts
86099 Unallocated Bal - I & DR

ORGANIZED ACTIVITIES
40113 Dairy Plant
40115 Farm
40126 Meats Laboratory
53107 Catering Laboratory
54116 Fine Arts
54131 Physical Education Activities
56131 Educational Communications
56199 Unallocated Bal - Org. Activities

EXTENSION AND PUBLIC SERVICE
71328 Nursery School
86352 Continuing Education

LIBRARY
86405 General Library
86431 Educational Communications
86499 Unallocated Bal - Lib.

STUDENT SERVICES
86501 Admissions
86516 Counseling
86520 Dean of Students
86530 Student Financial Aids
86531 Educational Communications
86542 Placement
86555 Registrar's Office
86560 Student Health Service
86570 Student Union
86599 Unallocated Bal - SS

MAINTENANCE AND OPERATION OF PLANT
86601 Administration & Supervision
86610 Custodial Service
86615 Maintenance: Buildings
86620 Maintenance: Equipment
86625 Maintenance: Grounds
86660 Security
86675 Utilities
86699 Unallocated Bal - M & O

GENERAL ADMINISTRATION
86701 President's Office
86731 Educational Communications
86738 Dean of Instruction
86741 Vice President for Administration
86765 Business office

86769 Campus Planning
86793 Purchasing
86799 Unallocated Bal - Gen Adm.

GENERAL INSTITUTIONAL SERVICES
86805 Automotive Services
86815 Central Duplicating & Printing
86820 Commencement
86825 Conference and Conventions
86835 Data Processing Services
86841 Educational Communications
86845 Mail and Messenger
86850 Membership Fees
86855 Public Relations
86865 Storehouse
86875 Telephone and Telegraph
86899 Unallocated Bal. - G.I.S.

AUXILIARY ENTERPRISES
87950 Residence Halls
87921 Food Service - FSA Contract
87931 Educational Communications
87966 Residence Hall Comm. Billing (Offset)

OTHER
98900 Savings
99999 Miscellaneous

Figure B-2   Sample list of charge numbers

Serial versus Shared Utilization

A second important problem concerning the recording of equipment utilization has to do with the fact that some machines operate on only one job at a time (serially), while others operate on two or more jobs simultaneously. The distinction between these two modes of equipment operation is perhaps most evident in the case of serial processing versus multi-programming or multi-processing computers.

In the case of a serial processing computer, which can only work on one job at a time, the recording of utilization information for costing purposes is quite straightforward. That is, a customer who uses the computer is billed for the use of the entire computer for the duration of the job. The fact that the customer's job does not require a particular tape drive or card reader is not important. What is important is that no one else can use any unused part of the computer while the job is being processed.

DATE _____

COMPUTER CENTER REQUEST

Director
MR. SEMPREVIVO   Phone 5337

REQUESTED BY _____PHONE _____

DEPARTMENT CHARGE CODE_____ REQUESTED COMPLETION DATE _____

JOB DESCRIPTION

# Copies	Preprinted Forms	Burst	Remove Carbons	Programmers Use Compile and Go	Test Compile
☐ ☐ ☐ ☐ 1 2 4 6	☐	☐	☐	☐	☐

FOR COMPUTER CENTER USE ONLY

Job Code _____ Job Name _____

Comments:	Machine	Time On	Time Off	Date Run
	29	_____	_____	_____
	59	_____	_____	_____
	83	_____	_____	_____
	85	_____	_____	_____
	514	_____	_____	_____
	557	_____	_____	_____
	360	_____	_____	_____
	000	_____	_____	_____
	001	_____	_____	_____
	002	_____	_____	_____

Figure B-3 Combined job request and machine utilization form

Operating on more than one job at a time as with multi-programming and multi-processing computers is quite a different case. Two or more jobs that share equipment during a given period of time must also share in the cost of that equipment for the same period of time. This method of sharing the cost of used and unused portions of equipment is frequently referred to as the "proration" of equipment costs.

To accurately prorate the costs of a multi-programming computer requires that the computer software possess the ability to keep utilization records. This "job accounting" or "job logging" capability, as it is sometimes called, enables the computer to automatically record how much of its capacity is being used by each job at any given time as well as the share of unused capacity which should be prorated to each job.

Any in-depth discussion of computerized job logging systems is beyond the scope of this appendix. Suffice it to say that job logging reports are an important requirement for determining how much computer time was used by each customer in a multi-programming environment.

Equipment Control—Lost Time

There is a third important control which must be employed if accurate equipment utilization data is to be made available. This is the maintenance of equipment time-lost records. Equipment time is considered to be lost whenever the equipment is not available for customer use during the time when it would normally be available for use. There may be many reasons for experiencing the loss of equipment time, such as equipment maintenance, malfunction, or inoperability. It is important to record the extent to which equipment is not available for customer use and the reason for the lost time. Actions taken by the machine operator or maintenance engineer which relate to problems causing the time loss should also be recorded. Figure B-4 illustrates the kind of information which might be recorded in a computer trouble log. Similar trouble sheets should be developed for all equipment and kept at a nearby location.

Another type of lost time which should be accounted for is commonly called "rerun" time. When a production job must be rerun because of an operator error, computer program error, machine malfunction, or other reason for which the customer is not responsible, it is common practice not to bill the customer directly for the original run time. The only chargeable service is the second or correct processing of the job. It is necessary to keep track of this kind of lost time to ensure that the customer is not improperly billed. Time lost

Date	Time	Equipment Description	Problem Description	Operator Action	Approx. Time Lost	Field Engineer's Comments

Figure B-4 Computer trouble log

because of reruns can be recorded into the regular computer trouble log by simply noting "rerun" beside the reason for the lost time.

Lost time should be eliminated from the analyst's determination of total equipment utilization since it represents time which was not truly available for customer use, hence it should not affect the cost of data processing services.

Computing Data Processing Equipment Costs

What we must now attempt to do is to summarize the equipment costs and equipment utilization so that an average rate can be established for each kind of equipment. Thus we should arrive at one billable rate for work done on our 029 keypunches, another for work done on our IBM 4341 computer, and still another for work done on our Univac 1100 computer. To do this, we must add to our basic equipment cost the following:

1. *Maintenance Cost*   This includes all maintenance repair and parts costs not covered in the basic equipment cost.
2. *Software Cost*   This includes all costs associated with operating systems and program products which the vendor has provided to us for a price.
3. *Expenses*   This includes all shipping charges and repair expenses for which the vendor has billed us.
4. *Supplies*   As noted in our discussion of supplies inventory control, this includes all supplies for which we have not already billed the customer. In arriving at our billable rate, we will actually have two charge rates—one for those who have been independently billed for supplies and one which includes a supplies cost factor.

By grouping these costs for a given type of machine (see figure B-5), we are able to isolate all costs associated with that type of equipment service. By dividing those costs by the total hours of productive machine utilization, we are able to arrive at the desired billable rate for equipment use:

$$\frac{\text{Equipment Costs}}{\text{Total Hours of Productive Equipment Utilization}} = \frac{\$10,000/\text{Month}}{400 \text{ Hrs./Month}} = \frac{\$25 \text{ Average}}{\text{Cost/Hour}}$$

Using this billable hourly rate and being able to identify each customer job, it is possible to determine how much to charge each customer for work which is performed or to estimate how much a particular new job will cost.

Summary of Monthly Costs for Equipment, Maintenance,
Software, Supplies, and Expenses for a Computer

Equipment rental/month	$ 7,771.00
Equipment maintenance/month	1,154.00
Software	700.00
Supplies and Expenses	375.00
Total	$10,000.00/month

Figure B-5   Monthly cost summary

Appendix C

Computer Evaluation and Selection

Developing the Concept of Workloads

From the discussion of task analysis in chapter 10, it was seen that a rather complicated operation could be reduced to its component parts or tasks and that these tasks could then be analyzed. If a truly representative workload sample which will assist in the evaluation and selection of computer systems is to be determined or developed, it will be necessary to apply this same approach to the analysis of total current and projected workloads.

The Analysis of Workloads

If one initially analyzes workloads by application area, it helps to maintain a degree of order and understanding to this process since many decisions regarding computer processing will be made by application area.

A workload analysis table (see figure C-1) can be developed for the entire workload. This table should include:

1. the name of the application area
2. the specific computer programs
3. the primary tasks involved in each program

In addition, it is important to note the individual run time (in hours) for each existing and proposed computer program and the frequency with which the program is run (average per month). By multiplying the individual run time by the number of times the program will be run in an average month, we are able to determine the average amount of computer time which a given program will consume in a given month.

However, knowing the above workload information at the computer program level is not sufficient to enable one to develop a representative workload sample. To derive this sample, it is necessary to break each computer program into its component tasks, such as: edit, sort, report, update, and compile. Notice, in figure C-1, that an attempt has been made to allocate portions of the overall program

Application and Program Tasks	Individual Run Time (Hours) (1)	Frequency Per Month (2)	Estimated Month Time (Hours) (1)×(2)=(3)	Sub-Totals by Task (4)
Application A				
Accounts Receivable				
A1: Recording of	.60	20	12.00	
Open Items				
a) Sort				10.00
b) Edit				1.00
c) Update				1.00
A2: Cash Application	.40	20	8.00	
a) Sort				5.00
b) Edit				1.00
c) Update				2.00
A3: Trial Balance	.75	20	15.00	
a) Report				15.00

Figure C-1   Sample Workload Analysis Table for Accounts Receivables Applications

run times to specific tasks. The precision with which this allocation can be done will depend on how sophisticated the existing computer time logging system is and on the way in which the program coding has been done. In any instance, it is an important step to be accomplished in developing a representative workload sample.

If we have done our work well, a completed workload analysis table will be available which indicates (by application area) all of the existing and proposed computer programs, their actual or estimated individual run times, and their average monthly run times. Furthermore, the task components for each program will be illustrated and allocated some portion of the overall program run time.

Summarizing Workloads by Task

Having obtained a breakdown of all individual program tasks, and a time factor for each task, it is now necessary to summarize these workloads by task and to translate task run time into a machine requirement. It is helpful to build a task summary for each task.

There are three parts to a task summary table. Figure C-2 illustrates how the input/output, or file description portion of the summary table might look if the total sort workload consisted of two sorts. Care has been taken to specify the kind and number of peripheral units required to perform each task. Whether the specific

	I/O or File Descriptions			
	Media Code	*No. of Devices*	*Device Category*	*Average Record Size (in char.)*
Task A1A (Sort)	PC	1	0	80
	MT	3	2	80
Task A2A (Sort)	PC	1	0	80
	MT	1	1	300
	MT	3	2	80

Code Key	
Media Code Explanation	*Device Category Explanation*
PC = Punch Card	0 = Source of original input
PT = Punch Tape	1 = Master File
MT = Magnetic Tape	2 = Intermediate, working or scratch
MD = Magnetic Disk	3 = Final Output
DD = Magnetic Drum	

Figure C-2   Sample workload analysis table for accounts receivable applications

peripheral unit is used as an input, output, working or master file has also been designated.

In figure C-3 we have expanded the table to include a second part—monthly average volume figures for transactions. We have also entered the estimated monthly task run times from our workload analysis table. (This transaction figure will be helpful later on when it becomes necessary to describe a typical sort program). Once this information has been entered into the table it is possible to construct the third and final portion of the task summary table which shows estimated equipment time requirements.

A sophisticated computer logging system will maintain an internal machine record of actual time usage for each device and possibly for the central processing unit. If your existing computer has such a logging system, this step in completing the table will be readily accomplished. If your current computer does not maintain such a log, then you are forced to estimate the amounts of time which peripheral devices are in use by the system.

When you do not have a complete computer log to work from it is impossible to estimate CPU time. However, an estimate of the CPU time is imbedded in the monthly task run time figure and is sufficient to enable us to continue.

Figure C-3 illustrates how part 3 of the task summary table may be completed for our limited example when a sophisticated computer logging system is not available. Once all entries have been made, totals are computed for the monthly task run time and individual peripheral run time requirements.

In constructing the task summary data, neither the internal storage requirement nor the programming languages to be used were specified. For the internal storage requirement, the minimum stated in the original equipment specifications serves as a guideline. That minimum reflects a real or estimated upper limit of need and is not as meaningful in working with a sample workload built upon average needs.

Programming language selections are based largely upon a user decision as to what languages best suit the needs and capabilities of the programming staff, as well as particular application requirements. Consequently, the languages to be employed in the performance test are quite simply dictated by the user after due consideration.

Selecting the Typical Program

Remember that it will be in the evaluation of systems capability that the upper limit requirements for systems hardware and software are explored. The performance evaluation will test a proposed computer's ability to perform against a representative or typical workload. The task summary table will help to determine what a typical task should look like. However, it must be clearly kept in

	I/O or File Descriptions				Monthly Averages	
	Media Code	No. of Devices	Device Category	Average Record Size (in char.)	Transaction Volume (1,000's)	Task Run Time (Hours)
Task A1A (Sort)	PC	1	0	80	200	10.00
	MT	3	2	80	15	
Task A2A (Sort)	PC	1	0	80	100	5.00
	MT	1	1	300	1	
	MT	3	2	80	3	

Figure C-3 Table expanded to include monthly average volume figures for transactions

	I/O or File Descriptions				Monthly Averages		Estimated Equipment Time		
	Media Code	No. of Devices	Device Category	Average Record Size (in char.)	Transaction Volume (1,000's)	Task Run Time (Hours)	Magnetic Tape	Card	Other
Task A1A (Sort)	PC	1	0	80	200	10.00	5.67	8.20	—
	MT	3	2	80	15				
Task A2A (Sort)	PC	1	0	80	100	5.00		4.10	—
	MT	1	1	300	1		1.00		
	MT	3	2	80	3		2.10		
					Total Sort Task Time	15.00	8.77	12.30	—

Figure C-4 Task summary table

mind that the amount of task run time each month weights the relative importance of that task's contribution to the typical workload. For example, in figure C-4, the total monthly run time for sort A1A is approximately twice the amount of run time for sort A2A. It can be assumed that the typical sort will be more like A1A than like A2A. It will be a weighted average rounded up to the nearest whole device and not a simple average of the two. The weighted characteristics of a typical sort task based on figure C-4 are:

Typical Sort Task

Media Code	No. of Devices	Device Category	Record Size	Transaction Volume (1000)
PC	1	Input	80	167.00
MT	1	Master	300	.33
MT	1	Work	80	11.00

Based upon the description of each of the typical tasks, the selection of a typical program can take either one of two alternative directions. One can look for existing programs which individually or together approximate the task description(s) in our original workload analysis table. Or, one can write a new program(s) which more exactly characterizes the typical workloads. To conserve time in an already lengthy process, existing programs should be used wherever possible.

Final Sample Workload Determination

Because we do not want to submit a typical program to a vendor and ask that it be processed for 15 hours, the final step in this process is accomplished mathematically. Figure C-5 illustrates how a transaction permits a small sample program to be an indicator of total workload requirements. Knowing the total monthly workloads for each task or program and for each device, it is possible to arrive at an extension (or weighting) factor. This factor indicates how many times the particular sample would have to be run if it were to equal the total monthly workload. Once we run the sample on the proposed computer system(s), we are able to multiply the proposed system run time by the factor(s) and then determine what the impact of the new system will be on the existing monthly workload.

Task Set	Workload Functions	Monthly Task	Representative Task (Single Run)	Extension Factor
Representative Programs *Time (Hours)*				
Sort	Total Thruput	145.00 ÷	0.45	322
B-1A	Magnetic Tape	125.00 ÷	0.25	500
	Card Reader	115.00 ÷	0.03	3,833
	Total Thruput	120.00 ÷	0.75	160
Edit	Magnetic Tape	80.00 ÷	0.60	133
E-4a	Card Reader	20.00 ÷	0.50	40
	Printer	100.00 ÷	0.25	400

Figure C-5 Final workload determination

Summary Comment

In this appendix an attempt has been made to outline in the briefest possible way how one might approach the following:

1. Analyze user workloads with an eye towards performance evaluation.
2. Organize workloads by component tasks and relate to equipment requirements.
3. Develop one or a series of typical programs which can function as benchmarks or as model components.
4. Extend individual programs as components of a total workload model.

Once an understanding of these basics is grasped, the student should be ready for an advanced study of performance evaluation techniques which extend beyond the scope of this text.

Index

Page numbers in **boldface** refer to the definitions in the Word Lists at the end of each chapter.

Accuracy, problems of, 130–31, **139**
Activity levels of jobs, 153–54
Actual key, 303, 306, 308, 309, **321**
Algorithm, randomizing, 306
American National Standards Institute (ANSI) flowchart symbols, 94–96
Assessment objective, 10
Assistance objective, 10
Audit requirements of systems, 264–68
Audit trail, 266–67, **285**
Automation, pros and cons of, 186–87

Backtracking, problem of, 158
Bar graphs, 222
BASIC, 350
Batch processing, 194–95, **208**
Benchmarks, 339, **352**
Benefits
  direct vs. indirect, 214, 216–17, 233
  intangible vs. tangible, 214, 215–16, 233
  recurring vs. nonrecurring, 214, 217, 233
  of system, defined, 214, 233, **234**
Billing statements, 232–33
Brainstorming, 67–68, **74**
Break-even analysis, 222–24, 233, **234**
Break-out points, 158
Business activity subsystem, 134–35, **139**, 292
Business function, 135–36, **139**, 292, 295
Business operation subsystem, 134, 136–37, **139**, 292, 294
Business systems
  characteristics of, 6
  collecting data about, 146–49
  defined, 5–6, **43**, 137–38, **139**, 292, 295
  exemplified by payroll system, 6–7
  unique elements in, 7–8

Calculation of data, 252
Card storage of data, 260
Cathode ray tube (*See* CRT)
Central processing unit (CPU), 327–28
Centralization vs. decentralization, 187–93
Charts
  of break-even analysis, 223
  Gantt, 329, 371, 372–74, **383**
  pie, 69–70
  military-style organization, 11–13, 23, **43**
  (*See also* Flowcharts)
COBOL, 314, 350
Communications
  asking effective questions for, 55–57
  attention to gestures during, 52–53
  defined, 10–11, **43**
  documentation and, 54–55
  essential for systems analysis, 10, 33, 49–50, 73
  in formal presentations, 69–72
  how to improve skills in, 72–73
  listening skills for, 51–52
  objectivity and, 53–54
  rapport basis for effective, 51, **74**
  training program in, outlined, 417–19
  using interviews in, 58–65
Computer Processing Manuals, 173–74
Computer programs
  application packages of, 339–40, 342
  benchmark, 339, **352**
  documentation for, 173–74
  estimating complexity of, 363–64
  utility, 342
  (*See also* Computer systems; Software)
Computer system evaluation and selection
  capability factor in, 334, 340–43

Computer system evaluation and
  selection (cont.)
  demonstration step in, 351
  of mini- and microcomputers, 349–
    51
  point-scale approach to, 346–49,
    352
  price/performance factor in, 334,
    336–40, **352**
  procedure for, 329–33, 351
  systems support factor in, 333, 334–
    36
  workloads analysis and, 432–38
Computer systems
  applications vs. modular approaches
    to, 196–99
  centralized, 188–89
  data communications and, 193
  distributive approach to, 191–92
  enhancement of, 327–28
  leasing, 337–38
  performance of, 328
  purchase plans for, 336–37
  real-time vs. batch processing in,
    193–95
  renting, 336
  stand-alone, 189–90
  star network, 190–91
  (*See also* Computer system
    evaluation and selection;
    Systems; Systems planning;
    Systems design)
Cost data
  availability of, 214
  categories of, 217–21
  comparative analysis of, 221–24
  on data processing, 224–33, 234
  life span of, 217
  on personnel work time, 226–29
  tangible vs. intangible, 218
Cost of services, 130
Costs
  comparative analysis of, 221–24
  of data processing, 224–33, 234
  direct vs. indirect, 214, 216–17,
    218, 233, **234**
  of equipment, 218–20, 221, 225,
    420–31
  of overhead, 220–21
  recurring vs. nonrecurring, 214,
    217, 218, 220, 233, **234**
  situational context and, 213–14,
    233
  of supplies and expenses, 220, 221,
    229–30
  of system, defined, 214, 215, 218,
    233, **234**
Costs/benefits, types of, 214–17, 233,
    234

Cost/responsibility centers, **43**, 231–
    32
Critical Path Method (CPM), 329,
    371, 374–81, **383**
CRT (cathode ray tube) screen design
  fill-in-the-blank approach to, 276
  HELP instruction approach to, 277–
    78
  to improve processing speed, 279
  input requirements for, 277
  need for, 275–76, 284
  transaction menu approach to, 278–
    79, **286**

Data
  administration of, 261–64
  area management as source of, 148,
    149
  area workers as source of, 149
  editing and correcting of, 249–52
  four levels of, 146
  issue of access to, 261
  organizing, 169–71
  outputting, 256–58
  processing, 252–56, 191–93, 208,
    328 (*See also* Data processing
    costs)
  raw, 146, **177**
  recording, 169–71
  reporting and documenting, 169–75
  retention of, 258–61
  storage media for, 259–61
  storing and updating, 252–56
  transmitting, 193–95, 208
  (*See also* Cost data; Data base;
    Data collecting and analysis;
    Data extraction and preparation;
    and other specific topics)
Data administrator, 261, 262, 264,
    285
Data banks, and security, 133
Data base (metabase)
  and data dictionaries, 264
  DBMS interface with, 313–14
  defined, 291–92, 297, 313, 320, **321**
  design of, 313–20, 321
  hierarchical model of, 316–17, 321,
    322
  network model of, 318–19, 321, **322**
  recovery procedures for, 406–8, 409,
    412
  relational model of, 319–20, 321,
    322
  schema and subschema of, 313–14
  top-down design of, recommended,
    315
Data base management system
    (DBMS)
  advancing development of, 320, 321

data dictionary compared to, 263–64

defined, 392, 297, 313, **321**

features of, 313–14

for system backup, 406, 408

three models of, 316–20

Data base recovery, 406, 408, 409, **412**

Data collection and analysis, 145–77

at area management level, 148, 149

in decision-making process, 68–69, 155

documentation process in, 169–75, 176

at job-performance level, 149

levels of information in, 146–49

pitfalls in, 150

recognizing problems during, 156–61, 176

as screening process, 146, 153

secondary sources of, 149

structured approach to, 150–53, 176

and systems planning, 20–23, 184, 185

tools used in, 22, 161–69, 176

what to look for on the job, 153–56

Data communications

defined, 193, **208**

impact of, 194–95

steps in, 193–94

Data description language (DDL), 314, 322

Data dictionary, 261–64, **285**

Data element, 261–63, 284, **285**, 293, 301

Data extraction and preparation

in conventional keypunch systems, 247–48

defined, **285**

design of source document for, 246

in online systems, 248–49

Data files

defined, 291, 320, **322**

direct-access organization of, 305–6

index sequential, 308–9, 321, **322**

inverted list structure, 311–12, 321, **322**

list structure, 309–10, 321, **322**

sequential, 304–5, 321, **323**

unordered, 309, 321, **323**

Data flow diagrams

advantages of, 88, 111

defined, **112**

examples of, 89, 242, 244

symbols used in, 88–90

Data manipulation language (DML), 314, **322**

Data processing costs

billing or estimating, 232–33

categories of, 224–25

of equipment, 225, 232, 420–31

hardware vs. software, 225

of personnel, 225–29, 232

of supplies, 229–30, 232

three-cost-center approach to, 230–32

Data records, 301–3, 316, **322**, 323

(*See also* Data files)

Data set (modem), 193, **208**

DBMS (*See* Data base management system)

Decision-making process, 68–69, **74**, 155, 156

Decision tables

applications for, 103

defined, 98, 111, **112**

examples of, 98, 100, 101, 102

multiple, 102–3

quadrant method of constructing, 99–101, **112**

results tables, 98–99, **112**

Detail sessions (of interview), 59, **74**

Directive interviewing, 64–65, **74**

Disk storage, 260

Distributive data processing (DDP)

defined, 191, **208**

pros and cons of, 192–93

Documentation process

defined, 169, **177**

illustrated, 170

special-purpose reports in, 171–73

steps in, 169–71

Systems documentation package

product of, 169, 173–74, **177**

value of, 174–75

DOUNTIL loop, 92

DOWHILE loop, 91

Earliest Start/ Earliest Finish Times (EST/EFT), 378, 380, 383

Economy, problems of, 130, **139**

Efficiency, problems of, 132, **139**

Environmental controls, 406

Equipment costs, 218–19, 221, 225, 420–31

Error-data test, 250

Exit session (of interview), 59–60, **74**

Extended Binary Coded Decimal Interchange Code (EBCDIC), 193

Feasibility

defined, 200, **209**

determining, 25–26, 171, 173, 213

four criteria for, 200–204, 208

preparing formal report on, 204, 205

Feedback, 4, **43**

Fields, 302

Fixed-format questionnaire, 103–10, **112**
Float time, 381, **384**
Flowcharts
    advantages of, 92
    in data collection and analysis, 161
    of data processing steps, 253
    examples of, 93, 162, 164, 165
    how to read, 97
    of organizational dynamics, 14
    of physical layout, 162–63
    process, 22, 161, 163–64
    program, 92, **112**, 174
    standard, 22, 92–97, 111, **112**
    symbols used in, 94–96, 163
    system, 92, **113**
    of updating process, 255
    mentioned, 72, 394
Forms, business
    control of, 269
    designing alternative types of, 272–74
    features of, 268–69
    input, requirements for, 270
    output, requirements for, 271
FORTRAN, 350
Free-format questionnaire, 103–4, **112**

Gantt charts, 329, 371, 372–74, **383**
Garbage In Garbage Out (GIGO), 250, **285**
Grandparent systems, 406, 407, **413**
Graphs, 70

Hard copy (of data), 259
Hardware
    costs of, 225
    documentation on, 346
    evaluating capability of, 328, 331, 334, 340, 343, 351
    minicomputers vs. microcomputers, 349–51, 352
    testing new, 394
HELP instruction, 278
HIPO (hierarchy plus input-process-output) technique, 72, 82–88, 110, **112,** 241, 280, 410

Informal organization, 11, **43**
Information, problems of, 133, **139**
Information activity, 300, **322**
Information bases
    activity in, 300, **322**
    data base, 291–92, 297, 320, **321**
    data file, 291, 320, **322**
    defined, 291, 320
    designing for, 252–56, 297–301, 320–21

file organization in, 303–11
    role of data records in, 301–3, **322**
    (*See also* Data base; Data file)
Information volatility, 300, **322**
Input requirements of systems, 246–52
Inputs, defined, 3
Interaction
    in functioning of system, 3–4, 6
    assessment of, by systems analyst, 10
Interview
    in data collection, 62–63
    defined, 58, **74**
    detail session(s) of, 59, **74**
    directive vs. nondirective, 64–65, **74**
    exit session of, 59–60, **74**
    handling resistance in, 60
    structured vs. nonstructured, 61, 62–64, **74**
    types and techniques of, 61–65
Introspection, and communication skills, **74**
Investment period, 223
IPO (input-processing-output) diagrams
    advantages and disadvantages of, 86–88
    defined, 110, **112**
    examples of, 87, 243, 245
    mentioned, 153, 244

Job-accounting system, 342, **352,** 428
Job activity, 153–54
Job auditing, 167, 168–69, **177**
Job delay, 159, **177**
Job dependence, 155
Job duplication, detecting, 157, **177**
Job effect, 155
Job inconsistency, 158, **177**
Job instruction, 161, **177**
Job logging (*See* Job-accounting system)
Job overlap, 157–58, **177**
Job sampling, 167, 168, **177**
Job significance, 154–55
Jobs, points of significance in analyzing, 153–56

Key-to-disk or key-to-tape recording, 248
Keypunching and verifying, 248
Keys, file
    actual, 303, 306, 308, 309, **321**
    link, 307
    logical vs. physical, 303
    primary, 303, 307, 309, **322**
    secondary, 303, 323

Latest Start/Latest Finish Times
(LST/LFT), 380–81, **383, 384**

Management
assessment of, in problem
definition, 120–22, 130
effect of, on systems analysis, 42–
43, 120–22, 184, 205
levels of, in MIS systems, 292–95
levels of responsibility in, 156
role of middle, or line, 148
screening data concerning, 146, 148
as source of data, 147, 148, 149
(*See also* Organizational structure;
Project management; Systems
department)
Management Information Systems
(MIS)
applications of, 296–97
computer-based, 295–96
defined, 292, 320, **322**
four levels of, 294–95
as pyramid of information, 293
Manipulation of data, 252
Meetings, techniques for effective,
66–67
Metabase (*See* Data base)
Microcomputers, 349–51
Microfiche, 259, **285**
Microfilm, 259–60, **285**
Military-style organization chart, 11–
13, 23, **43**
Minicomputers, 349–51
Missing-data test, 250
Models of systems organizations, 34–
42
Modem, 193, **208**
Modules, 197

Nassi-Shneiderman (N-S) charts, 90–
92, 111, **112**
Nominal group process, 68, **74**
Nondirective interviewing, 64–65, **74**
Nonstructured interviewing, 61, 62,
74

Open system, 3, **43**
Optical scanning, 248
Organizational commitment to
systems analysis
conflict and, 122–23
evaluating personnel for, 124
importance of, 205
role of management in, 42–43, 120–
22
steps in clarifying, 120
strategies for maintaining, 124–25
Organizational structure
flowchart of, 14
formal, 11

functional, 11, 16–17
hierarchical vs. functional analysis
of, 11–16
informal, 11, **43**
military-style chart of, 11–13, 23,
43
models for systems departments in,
34–42
relationship of systems analysis to,
32–43
span-of-control analysis of, 14–16,
44
Output requirements of systems, 256–
61
Outputs, defined, 3

Paper tape storage, 260
Parallel systems approach, 30
Paraphrasing, as communications
skill, **74**
Payback-period analysis, 222, 223–24,
233–34, **234**
Period of return, 223
Personnel
in systems implementation, 390
managing, 364–68
training of, 390–92, 393, 417–19
Pie charts, 69–70
Pilot or prototype systems, 203–4
Privacy, issue of, 405
Problem complexity
benefits of assessing, 134, 138–39
four levels of, 134–38, 292
Problem definition and classification
assessing problem complexity, 134–
38
benefits of, 138–39
essentials of, 119
gaining organizational commitment
as step in, 120–25
identifying kind of problem, 126–33
relationship of, to data collection,
118, 139
special-purpose reports for, 171–73
Problems (*See* Systems problems)
Processing requirements, 252–56
Program Evaluation and Review
Technique (PERT), 382, 383, 384
Program flowcharts, 92, **113**
Programming languages, 341–42, 350
Programs (*See* Computer programs)
Project management
project control in, 368–82, 383
responsibilities of, 357, 383
scheduling jobs and people in, 361–
68
skills for, 357–58, 383
task analysis in, 358–61, 383, **384**
Project manager, 357–58, **384**
Project Status Report, 369, 383

Quadrant method for decision table, 99–101, **112**
Questionnaires
  fixed-format, 103–4, 104–10, **112**
  examples of, 104, 105–6, 107
  free-format, 103–4, **112**
  guidelines for constructing, 105–8
  issuing, 108–9
  quantifying and interpreting, 109–10

Real-time processing, 194–95, **208**
Reasonableness test, for data, 250, **285**
Reliability, problem of, 131–32, **139**
Reports
  of data collection and analysis, 171–73
  on feasibility, 204, 205
  final systems proposals, 206–7, 331–32
  on project status, 369
Request for proposal (RFP), 330–31, **353**
Resource benefit, 201, **209**
Resource investment, 201, **209**
Resources, as feasibility criteria, 201–2
Responsiveness, problems of, 126–27, **140**
Responsibility levels, analyzing, 156
Results table, 98–99, **112**
Rules, in decision tables, 98

Schema of data base, 313, 315–21 *passim*, **323**, 405
Security, problems of, 132–33, **140**, 404–5
Self-correction in systems, 4–5, 6, **43**
Self-regulation in systems, 3, 5, 6, **43**
Set relationship (of data records), 316, **323**
Slack time, 381, **384**
Software
  costs of, 225
  documentation on, 346
  evaluating capability of, 328, 331, 334, 340, 341–42, 351
  DBMS, 296, 297, 314
  testing new, 394
  (*See also* Computer programs)
Span-of-control organizational analysis, 14–16, **44**
Stand-alone computing, 188–89
Standard flowcharts, 92–97, 111, **112**

Star network, 190–91
Statistical methods of analysis, 166–69
Steering Committee Model, 35–37, 41, **44**, 124–25
Storage media for data, 259–61, 306
Structured interviewing, 61, 62–64, **74**
Structured overview diagrams, 241
Structured, top-down analysis
  advantages of, 81, 239–45, 280, 283
  charting techniques used in, 81–92
  in data collection and analysis, 150–53, 176
  defined, 80–81, **112**
  example of HIPO technique in, 150–52, 153
  ideology underlying, 81
Structured walkthrough methodology, 281–83, **285**
Subsystems
  business activity as, 134–35, **139**, 292, 294
  business function as, 135–36, **139**, 292, 294
  business operation as, 134, 136–37, **139**, 292, 294
  defined, 5, **44**
Summary/status sheets, 169
Supplies, costing practices for, 220, 221, 229–30
Survey questionnaire (*See* Questionnaires)
System benefits (*See* Benefits)
System costs (*See* Costs)
Systems
  applications vs. modular approaches to, 196–99
  auditability of, 264–68
  backup and recovery methods for, 406–8, 409
  business, defined, 5
  centralized vs. decentralized, 187–93
  characteristics of, 3–5, 6
  conversion and implementation of, 28, 389–403
  data communications and, 193–95
  defined, 2, **44**
  determining costs and benefits of, 213–33
  determining feasibility of, 200–204, 208
  grandparent, 406, 407, **413**
  levels of, 5
  loop-like, 5
  mechanized vs. nonmechanized, 186–87
  methods for testing, 395–96

open, defined, 3, **43**
operating, 341
parallel, 30, 396–97, **413**
pilot or prototype, 27, 203–4, 400–402, **413**
preparing proposals for, 206–7
purpose of, 3, 6
quality assurance programs for, 403–12
review and evaluation of, 30–32
(*See also* Computer system evaluation and selection)
self-correction of, 4–5, 6, **43**
self-regulation of, 3, 5, 6, **43**
subsystems of, 5, **44,** 292
upgrading and downgrading, 327–28, **353**
(*See also* Computer system evaluation and selection; Computer systems; Systems design; Systems planning; and other specific topics)
Systems analysis
centralized vs. decentralized, 32
defined, 2, 8, 19, 42, **44**
evaluating team members for, 124
role of conflict in process of, 123, 138
standards for ideal arrangement for, 32
steps in, outlined, 19–32
and organizational structures, 32–42
use of decision tables in, 98–103
use of pilot systems in, 27
use of standard flowcharts in, 92–97
use of charting techniques in, 81–92
use of survey questionnaires in, 103–10
(*See also* Data collection and analysis; Problem definition and classification; Structured, top-down analysis; Systems design; Systems implementation; Systems planning; and other specific topics)
Systems analyst
contribution of, 8–9
communications training program for, 417–19
objectives of, 10–14
perspective of, 9, 11–17, 42
relationship of, to programmers, 280
requirements for becoming, 2, 42
role of, and technological growth, 9–10

Systems and Computers Evaluation and Review Technique (SCERT), 340, **353**
Systems benefits (*See* Benefits)
Systems checkpoints, 371–72
Systems conversion
defined, 389, **413**
dual systems method, 398–99, **412**
inventory method, 399, **413**
parallel systems method, 396–97, **413**
pilot systems method, 400–402, **413**
(*See also* Systems implementation)
Systems Departments
centralized vs. decentralized, 32
and documentation process, 169
models for, 34–42, 44
standards for ideal, 33
Systems design
advantages of structured approach to, 239–45
for auditability, 264–68
of business forms, 268–74
and completion schedule, 281
of CRT screens, 274–79
for data editing and correcting, 249–52
for data extraction, 246–49
defined, 28
for ease of use, 258
general considerations in, 199–200
of information bases, 291–321
logical (macro) phase of, 28, 238–39, 244, 283, **285**
physical (micro) phase of, 28, 238–39, 243, 283, **285**
for processing of data, 252–56
of program specifications, 279–81, 284
for quality control, 257–58
for retention of data, 258–61
for security, 286
use of structured walkthroughs in, 281–83, **285**
Systems Development Life Cycle
communications throughout, 50–51
defined, **44**
steps in, 19–32
Systems development team, 41–42, **44,** 47n
Systems documentation package, 169, 170, 173–74, **177**
Systems downgrading, 327–28, **353**
Systems feasibility (*See* Feasibility)
Systems flowcharts, 92, **113** (*See also* Flowcharts)
Systems implementation
continued by systems follow-up, 402–3

Systems implementation (cont.)
  defined, 389, **413**
  planning for, 389–96, 412
  simultaneous with conversion, 389
Systems Initiation Model, 34–35
Systems perspective, 9, 16–17, 42
Systems planning
  as continuous process, 184–85, 207
  considerations in, 183, 199–200
  determining feasibility in, 200–204,
    208
  final selection procedures in, 204–6
  formal proposal to conclude, 206–7,
    208
  major issues in, 186–99
  need for alternatives in, 185–86,
    208
  user and management involvement
    in, 184, 205, 207
Systems problems
  of accuracy, 130–31, **139**
  of economy, 130, **139**
  of efficiency, 132, **139**
  of information, 133, **139**
  of reliability, 131–32, **139**
  of responsiveness, 126–27, **140**
  of security, 132–33, **140**
  of throughput, 127–29, **140**
Systems proposals
  as contractual agreement, 205, 206,
    208
  elements of, 206–7
  purpose of, 26–27
Systems sign-off, 403, **413**
Systems upgrading, 327–28, **353**
Systems/user manual, 173

Tape storage, 260
Task analysis, 358–61, 383, **384**
Team Model, 41–42, **44,** 47n, 66–69,
    124–25, 242
Terminal data entry, 248
Test system migration, 408–10

Throughput, 127–29, **140**
Time and motion analysis, 167, 168,
    177
Transaction menu, 278–79, **286**
Transactional logging, 406, **413**
Translation of data, 252
Tuple, 319, **323**
Turnaround document, 274, 275, **286**

Unbundling, 346
Unreasonable-data test, 250, 285
Updating of data, 252
User involvement in systems
    analysis, 184, 200, 241–42, 283

Vendors of computer systems
  demonstrations by, 331
  negotiating with, 333
  notifying and debriefing, 332
  request for proposal to, 330–31, 343
  software packages by, 342
  support services of, 344–46, 352
  and system implementation, 393
  mentioned, 334, 336
Voice input, 248
Visual aids, 69–72
VTOC (visual table of contents)
    diagrams
  advantages of, 85–86
  defined, 110, **113**
  in data collection and analysis,
    150–53
  examples of, 82, 83, 84, 85
  how to write, 82–83
  revising, 84–85
  used with data flow diagram, 88

Workload distribution, 128–29, 159–
    60, **177**
Workload models, 339, 352, **353**
Workloads analysis, 329–30, 432–38
Workload simulation, 340, 352